Algorithms for Intelligent Systems

This book series publishes research on the analysis and development of algorithms for intelligent systems with their applications to various real world problems. It covers research related to autonomous agents, multi-agent systems, behavioral modeling, reinforcement learning, game theory, mechanism design, machine learning, meta-heuristic search, optimization, planning and scheduling, artificial neural networks, evolutionary computation, swarm intelligence and other algorithms for intelligent systems.

The book series includes recent advancements, modification and applications of the artificial neural networks, evolutionary computation, swarm intelligence, artificial immune systems, fuzzy system, autonomous and multi agent systems, machine learning and other intelligent systems related areas. The material will be beneficial for the graduate students, post-graduate students as well as the researchers who want a broader view of advances in algorithms for intelligent systems. The contents will also be useful to the researchers from other fields who have no knowledge of the power of intelligent systems, e.g. the researchers in the field of bioinformatics, biochemists, mechanical and chemical engineers, economists, musicians and medical practitioners.

The series publishes monographs, edited volumes, advanced textbooks and selected proceedings.

More information about this series at http://www.springer.com/series/16171

Shruti Jain · Meenakshi Sood · Sudip Paul
Editors

Advances in Computational Intelligence Techniques

Springer

Editors
Shruti Jain
Department of Electronics
and Communication Engineering
Jaypee University of Information
Technology
Solan, Himachal Pradesh, India

Meenakshi Sood
National Institute of Technical
Teachers Training and Research
Chandigarh, India

Sudip Paul
Department of Biomedical Engineering
North-Eastern Hill University
Shillong, Meghalaya, India

School of Computer Science
and Software Engineering
The University of Western Australia
Crawley, Perth, WA, Australia

ISSN 2524-7565 ISSN 2524-7573 (electronic)
Algorithms for Intelligent Systems
ISBN 978-981-15-2619-0 ISBN 978-981-15-2620-6 (eBook)
https://doi.org/10.1007/978-981-15-2620-6

This Springer imprint is published by the registered company Springer Nature Singapore Pte Ltd.
The registered company address is: 152 Beach Road, #21-01/04 Gateway East, Singapore 189721, Singapore

Preface

The aim of publishing the book is to serve for researchers, developers, and educators working in the area of recent advances and upcoming technologies utilizing computational intelligence in signal processing, computing, imaging science artificial intelligence, and their applications. As the book includes recent advances in research issues and applications, the contents will be beneficial to professors, research scholars, researchers, and engineers. This book will provide support and aid to the researchers involved in designing decision support systems that will permit the societal acceptance of ambient intelligence. Computational intelligence is a recently emerging area in fundamental and applied research, exploiting a number of advanced information processing technologies that mainly embody neural networks, fuzzy logic, and evolutionary computation. The book *Advances in Computational Intelligence Techniques* encompasses all branches of artificial intelligence which are based on computation at some levels such as artificial neural networks, evolutionary algorithms, and fuzzy systems. It presents the latest research being conducted on diverse topics in intelligence technologies with the goal of advancing knowledge and applications in this rapidly evolving field.

Accent from non-English speaking people creates a challenge in speech recognition and classification in systems based on English language. In order to tackle problems of recognition, systems should be trained in such a way that can understand accented voice of a particular accent targeting to specific geographical areas like Nigeria, Ghana, and Uganda, etc.

Electroencephalogram (EEG)-based brain–computer interface is applied in the classification of motor imagery EEG signal acquisition and analysis of data in clinical setup. This technique is having tremendous potential in reading the signal generated by different parts of our brain and providing the appropriate response to the specific limb. It focuses on feature extraction and feature selection for various activities. Such type of state-of-the-art technologies in signal processing may change the diagnosis and treatment process in several diseases.

Automatic glaucoma detection and assessment using fundus images are more dependent on segmentation features that may affect performance due to the chosen segmentation technique and feature extraction. Manual analysis and segmentation of fundus images are tough and time consuming. Convolutional neural networks (CNNs) are well known for their strong discriminative ability from difference intensity color images. The art of technologies like CNN in glaucoma diagnosis and localization can be more predictive and effective with higher diagnostic accuracy.

Diabetic retinopathy occurs due to high glucose level causing alterations in the retinal microvasculature.

Recent few years changed the support and living of disabled by increasing their self-dependency. Maximum of the problems are faced by the people who are having tetraplegic type of disability as they cannot move freely. For such type of people, technological intervention is much needed to increase their living standard and free movement. Assistive devices based on biological signals are playing a big role in upliftment and mobility freedom of this disabled community.

Heart rate variability (HRV) dynamics quantitative analysis based on fractal nonlinear analysis of R-R intervals provides a new approach for the assessment and treatment of attention-deficit tasks. The application of fractal or invariant scaling analysis to assess cardiac risk and forecasting can save many lives by making a novel model dealing with both hydrodynamic and homeostatic features.

Early detection and diagnosis of cervical cancer are the need of today's world as it is having an annual increment incidence rate. Early detection and accurate diagnosis can be made with the help of machine learning method reflecting effective and accurate analysis and classification.

Picture deblurring can enhance visual quality by reducing motion blur in dynamic visual inspection. This approach can be implemented with the help of deconvolutional algorithm-based model. This model works with deblurring and denoising by incorporating variable splitting. Deblurring calculation can be done by specifying saturated pixels of color images providing benefits in visual quality restoration and performance improvement.

In past few years, Internet of things (IoT) grafted its application in every field of our day-to-day life. Hospital administration and management have been changed a lot with the application of IoT and other new and modern technologies. IoT-supported systems are having enormous potential of information processing and data storing capability.

Melanoma (a type of skin cancer) diagnostics based on computer-aided texture segmentation is able to provide better quantitative and qualitative results. This can allow reproducible diagnosis leading to the decreased burden on physician by its inbuilt properties. This can be achieved by semi-automated and fully automated computer-aided diagnostic systems.

Machine learning, research, and development have flourished in the last few decades from general purpose to the specific purpose in our day-to-day life either directly or indirectly. It devotes to binary and arrays in data mining and pattern recognition. Its application can be seen in various fields like marketing, health care, travel and tourism, industry, and academia. One of its specific application that can be applied in the estimation of flight arrival with the help of directional antenna with adaptive beam-forming capability, spread spectrum receiver along with channel equalization. All these things can be done with the application of artificial neural network (ANN) and machine learning through support vector machine and direction of arrival estimation techniques.

Solan, India Shruti Jain
Chandigarh, India Meenakshi Sood
Shillong, India/Crawley, Perth, Australia Sudip Paul

Acknowledgements

At first, we would like to extend our gratitude to all the chapter authors for their sincere and timely support to make this book a grand success.

We are equally thankful to all executive board members of Springer Nature for their kind approval and for granting permission to us as editors of this book.

We would like to extend our sincere thanks to Mr. Aninda Bose, Senior Editor, Hardsciences, Springer Nature, and Raashmi Ramasubramanian (Ms.), Production Coordinator (Books) and Silky Abhay Sinha (Ms.), Project Coordinator, Books Production, Springer Nature, for their valuable suggestions and encouragement throughout project.

With immense pleasure, we express our thankfulness to our colleagues for their support, love, and motivation in all our efforts during this project.

We are grateful to all the reviewers for their timely review and consent, which helped us lot to improve the quality of book.

There are so many others whom we may have inadvertently left out, and we sincerely thank all of them for their help.

Contents

About the Editors

Dr. Shruti Jain is Associate Professor in the Department of Electronics and Communication Engineering at Jaypee University of Information Technology, Waknaghat, H.P, India and has received her Ph.D. in Biomedical Image Processing. She has a teaching experience of around 15 years. Her research interests are Soft Computing, Image and Signal Processing, Bio-inspired Computing and Computer-Aided Design of FPGA and VLSI circuits. She has published more than 07 book chapters, 60 papers in reputed journals, and 40 papers in International conferences. She has also published five books. She is a senior member of IEEE, life member and Editor in Chief of Biomedical Engineering Society of India and a member of IAENG. She has completed one externally funded project and one in the pipeline. She has guided 01 Ph.D. student and now has 06 registered students. She is a member of the Editorial Board of many reputed journals. She is also a reviewer of many journals and a member of TPC of different conferences. She was awarded by Nation Builder Award in 2018-19.

Dr. Meenakshi Sood is Associate Professor in National Institute of Technical Teachers Training & Research, Chandigarh, MHRD, GOI, India She is Gold medalist in her M.E (ECE) and holds a Ph.D. in Biomedical Signal Processing. Her research interests are in Bio-inspired Computing, Image and Signal Processing, Antenna design, Metamaterials and Soft Computing. She has published more than 50 papers in reputed journals and 60 papers in international conference proceedings. She is a senior member of the IEEE and life member of the Institute of Engineers, Biomedical Engineering Society of India, and a member of the IAENG. She has two externally funded projects and has another one in the pipeline. She has published two books for undergraduates.

Dr. Sudip Paul is currently an Assistant Professor in Department of Biomedical Engineering, School of Technology, North-Eastern Hill University (NEHU), Shillong, India since 2012. He completed his Post doctoral research at School of Computer Science and Software Engineering, The University of Western Australia, Perth. He was one of the most precious fellowship awardee (Biotechnology

Overseas Associateship for the Scientists working in North Eastern States of India: 2017-18 supported by Department of Department of Biotechnology, Government of India). Apart from this affiliation, he is also as. He received his Ph.D degree from Indian Institute of Technology (Banaras Hindu University), Varanasi with specialization in Electrophysiology and brain signal analysis. He has many credentials in his credit out of which his First Prize in Sushruta Innovation Award 2011 sponsored by Department of Science and Technology, Govt. of India and also he also organized many workshops and conferences out of which most significant are the 29th Annual meeting of the Society for Neurochemistry, India and IRBO/APRC Associate School 2017. Dr. Sudip published more than 90 International journal and conference papers and also filled four patents. Recently he completed three book projects and two are ongoing as editor. Dr. Sudip is a member of different Societies and professional bodies, including APSN, ISN, IBRO, SNCI, SfN, IEEE and many more since 2009, 2010, and 2011 onwards. He received many awards specially World federation of Neurology (WFN) traveling fellowship, Young Investigator Award, IBRO Travel Awardee and ISN Travel Awardee. Dr. Sudip also contributed his knowledge in different international journals as editorial board members. He has presented his research accomplishments in USA, Greece, France, South Africa and Australia.

Chapter 1
Accent Classification of the Three Major Nigerian Indigenous Languages Using 1D CNN LSTM Network Model

Ayodeji Olalekan Salau, Tilewa David Olowoyo and Solomon Oluwole Akinola

1 Introduction

Research interest in recognizing and synthesizing accents dates back to several decades ago, but just until the mid-twentieth century that the automatic speech recognition (ASR) system was invented [1]. ASR systems are useful for the identification of spoken languages, especially in natural language processing (NLP). NLP is a subbranch of artificial intelligence (AI) that requires computational technologies and linguistics to process natural languages like humans. Although there are some NLP technologies available today such as Siri from Apple and Alexa from Google [2], these technologies are not used to identify and differentiate between different accents of various dialects [3]. The concept of NLP has been applied to different aspects of research and technological development such as text classification, image classification, and speech recognition.

In natural language processing, accent identification and classification is a major challenge [4]. AC is a method of determining a persons accent by taking speech from speakers of various ethnic groups or nationalities and using it to determine the speaker's background or nationality. This provides a way of uniquely extracting useful information of where the person is from without necessarily asking the person, but instead by just listening to them speak. These and many more, still show that speaker recognition systems are still not able to match the performance of human recognition systems.

The rest of this paper is organized as follows. Section 2 presents a review of related works. Our proposed method is introduced in Sect. 3. The experimental results are presented and discussed in Sect. 4, and Sect. 5 concludes the paper.

A. O. Salau (✉) · T. D. Olowoyo · S. O. Akinola
Department of Electrical/Electronics and Computer Engineering, Afe Babalola University, Ado-Ekiti, Nigeria

S. Jain et al. (eds.), *Advances in Computational Intelligence Techniques*, Algorithms for Intelligent Systems,
https://doi.org/10.1007/978-981-15-2620-6_1

2 Related Works

In various parts of the world, different races communicate with each other using their individual dialects. In [5], Australian accents were classified using the mel-frequency cepstral coefficients (MFCCs) to recognize and extract various accents from speech. The Australian accent is classified into three major varieties of spoken English as cited by linguist, namely: General (55%), spoken (34%) and cultivated (11%). The authors used speech information called voice signatures such as age, gender, and emotion to classify Australian accent. These voice signatures are used in the design of APR systems and speech mining applications.

A text-dependent approach was proposed in [6] using Gaussian mixture model (GMM) technique for automatic accent identification (AID) to identify accents among four regional Indian accents with speakers gender. In [7], MFCC-GMM-based approach was used for accent recognition of Indian Telugu speech signals. An accuracy of 91% was achieved for speaker recognition. A neural network approach was used in [8] for Persian accent identification. The authors achieved a high result of 98.51% for training of Kermanshahi accent and 97.75% was achieved for testing.

In [9], identification of Indian (Hindi) accent was achieved. Emotions such as fear, anger, sadness, happiness, and disgust were used to distinguish between the different accents. Behravan et al. [10] analyzed the factors affecting foreign accent recognition using Finnish accent as a case study. This time the authors used spectral features from the *i*-vector system on Finnish foreign accent recognition task. The results obtained show that accent identification is easier for older speakers than it is for younger speakers. A comparison of different classifiers was achieved in [11] for speaker accent recognition. The classifiers compared with respect to computational time and accuracy are support vector machine (SVM), k-nearest neighbor (kNN), linear discriminant analysis (LDA), quadratic discriminant analysis (QDA), and MFCC. Among these classifiers, the kNNs was found to perform better than the rest. In [12], a phone recognition followed by language modeling (PRLM) approach was used for identification of Arabic dialect. Four colloquial Arabic dialects, namely: Iraqi, Egyptian, Gulf, and Levantine were tested using the proposed approach.

2.1 Speech Science and Feature Extraction

Speech is analog in nature and can be represented as a two-dimensional vector made up of samples and a corresponding amplitude. Speech is produced by a speaker in the form of sound wave which travels from the speaker to the listener's ear. MFCC is the most popular feature extraction (FE) method used for AC for identification of acoustic features of a speaker [13, 14]. Accent is described as a feature of pronunciation which identifies a person's social and regional background or characterizes an individual's speech as belonging to a particular ethnic group [15], while a dialect is any distinguishable variety of a language spoken by a group of people likened to their grammar and vocabulary. Equation (1) gives the formula for the conversion of linear scale frequency (LSF) to mel scale frequency (MSF), while Eq. (2) gives the conversion of MSF to LSF.

$$f_{\text{mel}} = 2595\ \log_{10}(1 + f/700) \tag{1}$$

$$\text{M}^{-1}(\text{m}) = 700\,(\text{Exp}(\text{m}/1125) - 1) \tag{2}$$

where f_{mel} and f are the mel and linear frequencies, respectively.

2.2 Accent Features

Perceived pitch or frequency of tone is measured in mel. Mel scale is approximately linear below 1 kHz and logarithmic above 1 kHz. The accent features which are commonly extracted are MFCCs, pitch, intonation, energy, formants, and formants trajectories.

In [16], prosodic features such as syllables of words, syllable duration, intensity, and F0 temporal variation were used for foreign accent identification. Hidden Markov Model (HMM) was further used for training each word and accent. The likelihood of accent identification gave an estimate of 16 Gaussians with HMM. Experimental results show a correct identification rate of foreign accents of Italian, French, Greek, and Spanish speakers of 87.1% using MFCC. Long and short features were used for accent identification in [17] by combining deep neural networks (DNNs) and recurrent neural networks (RNNs) on Italian, Korean, Arabic, and Japanese accents.

2.3 Mel-Frequency Cepstral Coefficients Extraction Technique

Feature extraction (FE) is a technique used to identify the components of an audio signal that are suitable for extracting its linguistic content [13]. It is important to note that the shape of the vocal tract filters the sounds generated by a human being and this shape determines what sound comes out. If the shape of the vocal tract is determined accurately, then it gives an accurate representation of the phoneme being produced. The shape of the vocal tract manifests itself in an envelope of the short-time power spectrum, and the job of MFCCs is to accurately represent this envelope. Prior to the introduction of MFCCs, linear prediction coefficients (LPCs), and linear prediction cepstral coefficients (LPCCs) were used in ASR.

Nowadays, MFCCs is the FE technique which is most widely used in ASR. MFCCs gives the Fourier transform of a signal which map the power spectra obtained as a mel scale using a triangular overlapping window and takes the log power of each mel frequency along a discrete cosine transform (DCT) of the mel log returning the amplitude of the resulting spectrum of signal.

2.4 Challenges with Nigerian-Accented English

Quite a handful of authors have worked on accent classification using Nigerian-Accented English (NAE), even as NAE has some common phonological properties which can be used to identify it despite its subvarieties. In [18], a comparative study of Nigerian English (NE) with American English (AE) was achieved. The study focused on identifying sources of mismatch between NE and AE by using acoustic signal modeling and evaluation. The evaluated models gave a reduction of ASR errors of 37% absolute word error rate.

Based on our review of related works, this paper presents an accent classification method for Nigerian-Accented English (NAE) using the major three Nigerian indigenous languages (Hausa, Igbo, Yoruba) by applying MFCC for FE and 1D CNN LSTM for training and testing the dataset. Although the CNN LSTM architecture is specifically designed for problems with spatial inputs, like images or videos, in this work, we have used it for speech signals.

3 Proposed Methodology

This section discusses the steps that were taken in developing the accent classification method using one-dimensional convolutional neural network (1D CNN) and long short-term memory (LSTM) network model (1D CNN LSTM). The method overview is shown in Fig. 1.

3.1 Audio Data Acquisition and Preprocessing

Audio data in the form of speech were collected from 100 speakers from each of the three major Nigerian indigenous languages, namely: Hausa, Igbo, and Yoruba. The

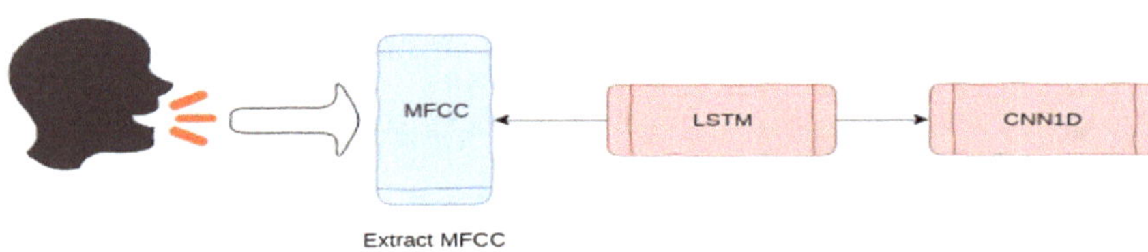

Fig. 1 Overview of the proposed method

speakers from these three major Nigerian languages were given a list of 20 different words to pronounce, making a total of 300 speakers and 6000 utterances. The words were pronounced within a period of 20–25 s. The chosen words are words which these respective groups (tribes) find difficult to pronounce properly as compared to English native speakers. They were chosen because of their ability to bring out the wanted accents from the speakers.

The speech data were acquired using a mobile phone as a recorder. Recordings were taken at a sampling rate of 20,000 Hz in a serene environment with low background noise. The acquired speech data consists of continuous telephone-quality speech in English from the different languages stored in WAV format.

The speakers were made to pronounce words such as: Direct, House, Church, Drive, Pear, Horse, Room, Champion, Machine, Torch, Problem, Proud, Rag, Happy, Bright, Chance, Liver, Fry, Grab, and Lag.

The acquired speech data were preprocessed and features were extracted using MFCC to give an input shape of (40, 1500) six LSTM layers. The LSTM layer was used with a batch normalization function which helped to normalize the data by performing the mean and standard deviation of the acquired speech data.

3.2 *Audacity*

Audacity software was installed in the recording device. This software is an open-source digital audio editor and recording application for Windows and other operating systems. Apart from recording audio from multiple sources, Audacity is also used for post-processing of all types of audio data which are imported into it. Audacity's noise reduction effect can be used to clear background noise such as noise from electric fans or hums.

After we saved all the acquired speech data into a folder which we named "acquired speech data," the data were imported into the Audacity software to remove background noise and after that, the speech data was exported from the mp3 format into the .wav format. This format is a less compressed format suitable for audio processing.

3.3 *Feature Extraction*

This section gives a brief description of the steps taken for FE using MFCCs. FE was performed by reading the audio speech into a numerical vector with librosa library which helps to convert the speech data to input MFCC function of librosa to compute the MFCC vector. The MFCC vector obtained was pad with zero to replace the part in the audio amplitude where there was no signal with numpy. The values of the MFCC vector form an array that was declared X. The target variable was also extracted from the data loader and then forms an array of Y.

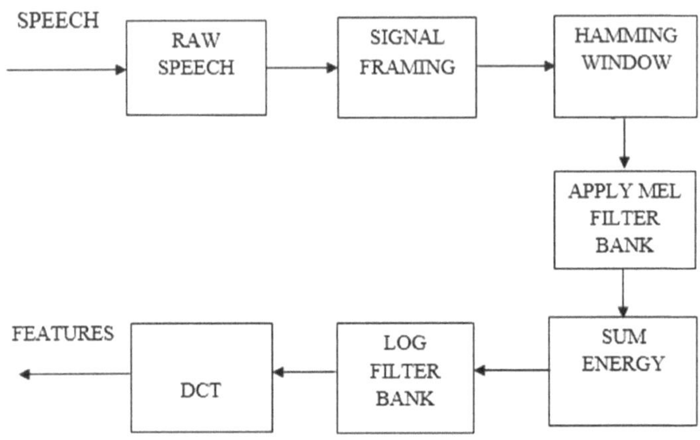

Fig. 2 MFCC block diagram

MFCC comprises of a mel scale which is used to relate perceived frequencies of pure tones to their actual measured frequency. This was achieved with the following steps as shown in Fig. 2.

(a) Subdivide the signal into short frames.
(b) For each of the frames, calculate the periodogram estimate of the power spectrum.
(c) Employ the mel filter bank to the power spectra, and add each filters energy.
(d) Obtain the logarithm of all filter bank energies.
(e) Obtain the DCT of the log filter bank energies.

The conversions are performed with Eqs. (1) and (2).

3.4 Accent Classification Algorithm

Speech data from the three major Nigerian indigenous languages were grouped into Hausa, Yoruba, and Igbo speakers. Table 1 presents the accent classification algorithm based on speaker groups. The flowchart of our proposed accent classification method is shown in Fig. 3.

The proposed 1D CNN LSTM network model consists of six LSTM layers of which each layer contains different unit values which are normalized using batch normalization function (BNF) as shown in Fig. 4. BNF helps to normalize each output of the models layers.

Table 1 Speaker classification algorithm

Accent Classification Algorithm
Requirement: Speech data
Ensure speech signals in dataset are Hausa, Igbo, and Yoruba accents
Read dataset [igbo, yoruba, hausa]
define mfcc (file, audio length)
read and covert (audio, audio_length)
pad audio with zero
return audio array
define train data (audio)
mfcc (audio)
get label from audio path append to array y, get mfcc data and push to array x
return train x and y
define test data (audio)
mfcc (audio)
get label from audio path append to array y, get mfcc data and push to array x
return test x and y or validation
convert label to integer 1, 2, 3, and to binary encoding, number of class = 3
callback Model checkpoint; save the best model on the validation side
define Sequential model
add LSTM (200, input_shape = (40, 1500), return_sequences = True)
add LSTM (100 unit, return_sequences = True)
add LSTM (50 unit, return_sequences = True)
add Batch Normalization
add LSTM (35 unit, return_sequences = True)
add LSTM (20 unit, return_sequences = True)
add Batch Normalization
LSTM (18 unit, return_sequences = True),
add Conv1D (16, activation = relu, kernel_size = 3, padding = 'same')
add MaxPool1D (pool_size = 2), add Batch Normalization
Flatten the array to 1-dimension
add Dense (60 neuron, activation function = relu),
add Dense (5, activation function = relu),
add Dense (number of class, activation function = softmax)
loss category_cross entropy, optimization algorithm (Adam), Metric Accuracy
compile model (loss, optimizer, metric)
Train model (X, Y, epoch = 200, validate (x_test, y_test), callback)
end

- The first layer has input value of (40, 1500) and output of (40, 1500).
- The second layer consists of two LSTM which are the input values. First input (40, 1500) gives an output (40, 200). The second input (40, 200) gives (40, 100) because three LSTM were stack together and were normalize with batch norm to give an output value of (40, 50).
- Third layer takes the output of the batch normalization and reduces it to (40, 35). This value is inputted into another LSTM which is stacked together and the output is (40, 20). This value is normalized to give an output of (40, 20).
- Another LSTM layer takes in (40, 20) of data and outputs (40, 18) which is stacked on a convo 1d which outputs (40, 16) of data.

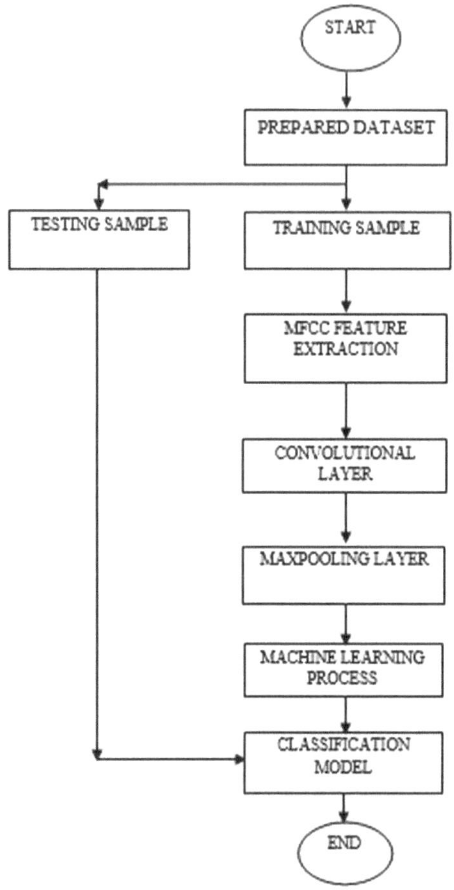

Fig. 3 Flowchart of the proposed accent classification process

- Max pooling 1d was done to pull the high values. This reduces the pool value to (20, 16) which was flattened to 320. This linear value was inputted into a dense layer which is known as artificial neural network (ANN) layer.
- The dense layers are three with input 320 each. They later reduce to an output of 60 for the next layer which then takes the 60 and reduces it to 5.
- The last layer inputs this 5 and outputs a value of 3 which is the predicted value.

The activation function used on the LSTM is tanh while on the convolution 1d relu activation function was used. Similarly, the dense layer used relu for dense_1 and dense_2 layers, and the final layer used softmax to output the probability of it being the class. 1D CNN is employed in this work because it is most preferred for audio signal analysis and processing as compared to the 2D CNN which is mostly used for images, and 3D CNN which is used for movies.

Fig. 4 1D CNN LSTM Model of the accent classification procedure

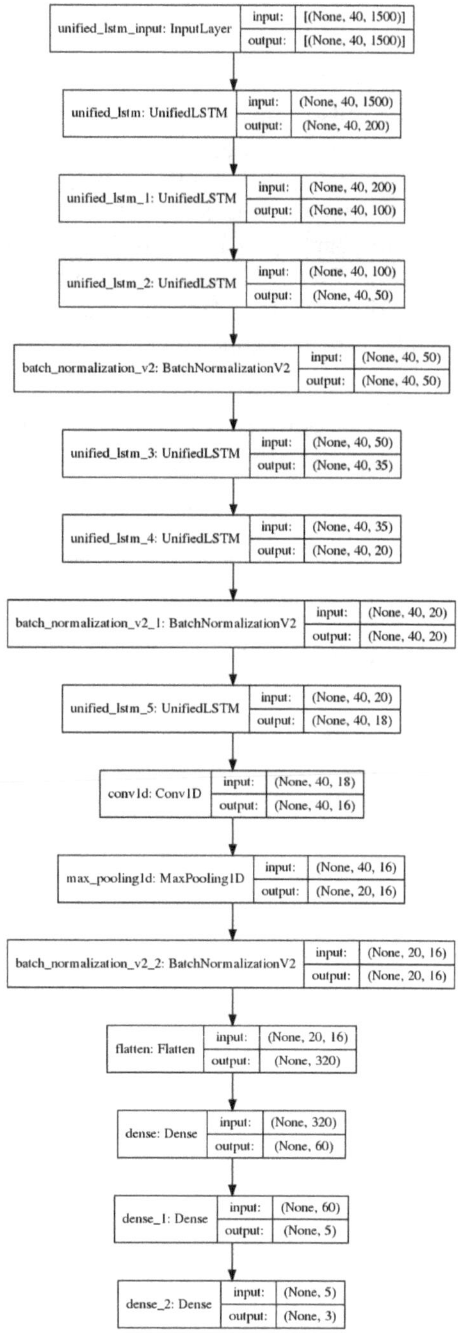

4 Results and Discussion

In this section, we present the experimental results for training and testing. In addition, a comparative evaluation of our results in terms of the accents, classification accuracy and classifiers used are shown in Tables 5 and 6, respectively. For the purpose of experimentation, we have evaluated our proposed approach using an Intel Core i5 PC with 2.6 GHz CPU, 8 GB RAM, and Windows 8 operating system. Python programming language was used on TensorFlow 2.0 Python development environment. The major Python libraries used were: librosa, tkinter, numpy. Matplotlab, panda, and scikit learn scipy. The audio signal processing, audio noise removal, training and testing stages are discussed in Sects. 4.1, 4.2, and 4.3, respectively.

Table 2 The model performance for the three different accents

Model approach (1D CNN LSTM)	Hausa accent (%)	Igbo accent (%)	Yoruba accent (%)	Average (%)
Training accuracy	98	91.5	97.2	95.6
Test accuracy	97.7	90.6	96.4	94.9

Table 3 Results of values obtained from features

Features	Value
Data format	.wav
Mode	Mono
Audio rate	20,000
n_mfcc	40

Table 4 Functions and their parameters

Function	Parameters
6 stack LSTM	[200, 100, 50, 35, 20, 18] unit
Stack 2 neural network	[60, 5, 3] values of neurons
Batch size	32
Epoch	120
Loss	Categorical cross entropy for multiclass
Optimizer	Adams
Learning rate	0.002
Batch normalization	3

Table 5 Comparison of results of the classification accuracy for various countries

Author	Dialect/accent	Country	Accuracy (%)
[10]	Finnish	Finland	Not specified
[17]	Arabic, Italian, Korean, and Japanese	Arabic, Italian, Korean, and Japanese	42.35, 68.08, 47.78, and 44.71
[22]	Karnataka (Mumbai Mysore, Karavali)	India	76.7
[9]	Hindi	India	81
[25]	Chinese	China	86
[24]	Malaysian, Chinese, and Indians accent	Malaysia	87.97
[5]	Australian	Australia	90
[7]	Telugu	India	91
[6]	Kashmiri, Manipuri, Bengali, and neutral Hindi	India	97, 76, 68, and 89
[12]	Arabic	Levantine, Egyptian, Gulf, Iraqi, and MSA	97
Proposed approach	Hausa, Igbo, and Yoruba	Nigeria	97.7, 90.6, and 96.4

4.1 Audio Signal Processing

Speech data were acquired using a phone as the recording device at a sampling rate of 20,000 Hz. Precisely, 6000 audio samples were collected. In the course of data collection, it was observed that the collected speech data samples were mixed with lots of background noise emanating from the surrounding environment. This noise was filtered out by analyzing the acquired speech data using Audacity.

4.2 Noise Removal Using Audacity

The results of the acquired speech data with and without noise are shown in Figs. 5 and 6. The results in Fig. 5 suggest that the acquired data was found to contain little background noise, while that of Fig. 6 shows that Audicity was effective in removing noise from the acquired speech data.

Table 6 Comparison of features extracted and techniques employed

Author	Features	Classifier type
[4]	Prosodic features	Neuro fuzzy classifier
[5]	MFCC	GMM
[6]	MFCC	GMM
[8]	MFCC	SVM and kNN
[9]	MFCC	SVM and Auto-associative neural network (AANN)
[10]	SDC + MFCC	i-Vector based
[12]	MFCC and delta features	PRLM and ergodic HMM
[17]	Long and short features	DNNs and RNNs
[19]	Spectral and prosodic features	SVM and an ensemble of decision tree
[20]	MFCC	GMM and SVM
[21]	Delta–delta mel-frequency cepstral coefficients (DDMFCC)	GMM
[22]	MFCC	GMM
[23]	MFCC	GMM and HMM
[24]	MFCC	kNN
[25]	MFCC	GMM
Proposed approach	MFCC	1D CNN and LSTM (1D CNN LSTM)

4.3 Training and Testing

This section reports the results obtained from the experiments performed using our proposed method. The speech data were trained using the 1D CNN LSTM network model with Python neural networks library with Keras running on the TensorFlow 2.0 Python development environment. The training and testing results are presented in Table 4.

The results show an average accuracy of 95.6% for training and 94.9% for testing. Furthermore, Hausa accent outperformed the other two accents in both its training and testing accuracy. This was followed by the Yoruba accent and lastly, the Igbo accent. The Adams algorithm was used for optimization. A high learning rate of 0.002 was achieved with the Adams algorithm as shown in Table 5. In addition, we have used a batch normalization of three to speed up the training time. The max pooling layer retains important information about the audio speech data while it reduces the dimensionality of the input and thus the computations in the network. The layers of the 1D CNN are densely connected with the last layer using a softmax layer to output the confidence of each class predication. The model accuracy and model loss at 120 epoch are shown in Figs. 7 and 8, respectively. Fig. 7 shows that a higher training accuracy was achieved compared with its testing accuracy with both

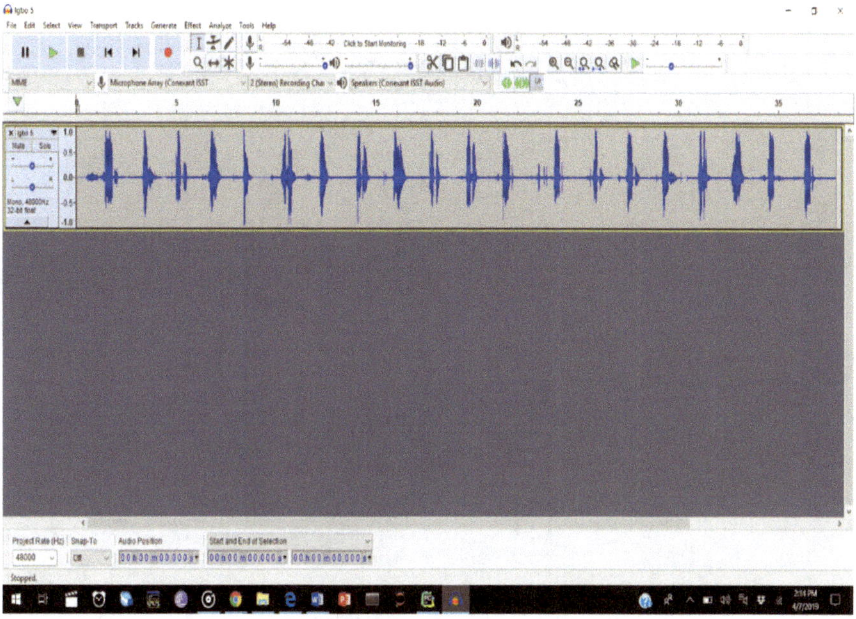

Fig. 5 Audio recording with background noise

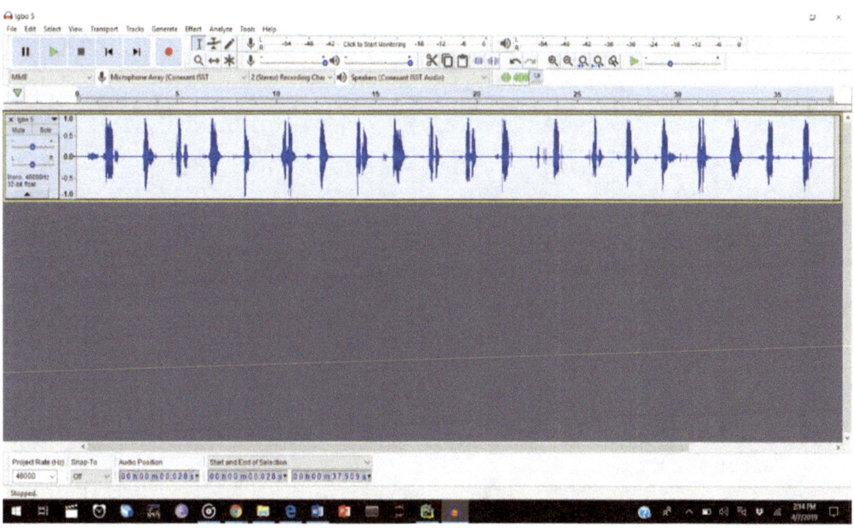

Fig. 6 Audio recording without background noise

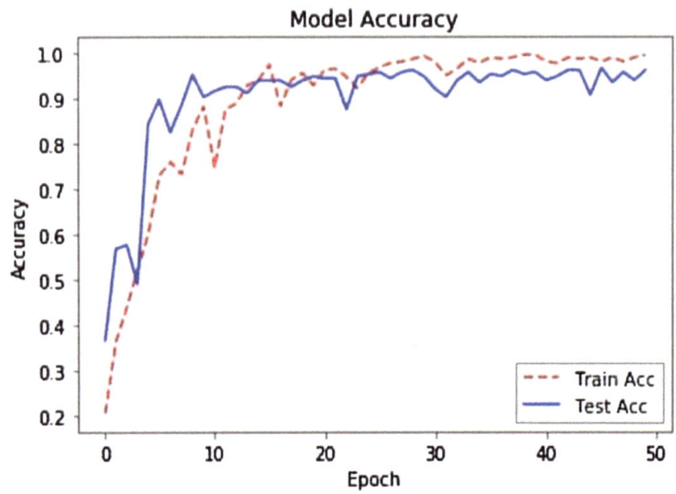

Fig. 7 Graph of model accuracy

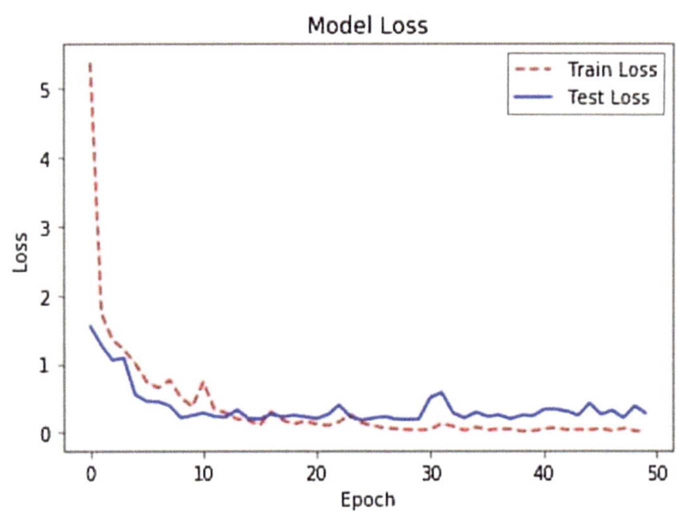

Fig. 8 Graph of model loss

having similarities at the initial stage of simulation. In Fig. 8, a lower loss rate was achieved for the proposed model with a consequently lower loss rate for training than testing. The vertical scale represents the accuracy with 100% representing 0.1 and the horizontal scale representing the speech epoch. In Table 6, we show the result of values obtained from features used. The.wave format is the original format saved in the sound file which shows the original waveform of the sound signal collected over the sampling period.

In Table 2, we compare our results with existing works. The comparison of results shows that the proposed 1D CNN LSTM network model is seen to outperform other machine learning techniques used for accent classification for different countries.

Table 3 shows a comparison of the different types of features extracted and the classifier types using the proposed 1D CNN LSTM network model as compared to other research papers.

5 Conclusion

Accent classification of the three major Nigerian indigenous languages was achieved in this work. The results show the high capability and accuracy of the 1D CNN LSTM network model to identify and classify native NAE speakers into the three major indigenous Nigerian languages. MFCC features were extracted from the acquired speech data recordings. The proposed 1D CNN LSTM network model with the designed algorithm was able to perform classification of speakers into Hausa, Igbo, and Yoruba giving an average accuracy of 97.7%, 90.6%, and 96.4%, respectively. In the future, we hope to use a more complex neural network architecture on a larger dataset and probably fuse a number of techniques together to get a better performance.

References

1. Juan SFS (2015) Exploiting resources from closely-related languages for automatic speech recognition in low-resource languages from Malaysia. Ph.D. Dissertation, Université De Grenoble, pp 1–146
2. Matthew B, Hoy MB (2018) Alexa, Siri, Cortana, and more: an introduction to voice assistants. Med Ref Serv Q 37(1):81–88. https://doi.org/10.1080/02763869.2018.1404391
3. Lulu L, Elnagar A (2018) Automatic arabic dialect classification using deep learning models. Procedia Comput Sci 142:262–269. https://doi.org/10.1016/j.procs.2018.10.489
4. Sarma M, Sarma KK (2016) Dialect identification from Assamese speech using prosodic features and a neuro fuzzy classifier. In: 3rd international conference on signal processing and integrated networks (SPIN). Noida, pp 127–132. https://doi.org/10.1109/spin.2016.7566675
5. Nguyen P, Tran D, Huang X, Sharma D (2010) Australian accent-based speaker classification. In: IEEE third international conference on knowledge discovery and data mining, pp 416–419. https://doi.org/10.1109/wkdd.2010.80
6. Malhotra K, Khosla A (2008) Automatic identification of gender and accent in spoken Hindi utterances with regional Indian accents. In: 2008 IEEE spoken language technology workshop. Goa, India. https://doi.org/10.1109/slt.2008.4777902
7. Mannepalli K, Sastry PN, Suman M (2016) MFCC-GMM based accent recognition system for Telugu speech signals. Int J Speech Technol 19(1):87–93. https://doi.org/10.1007/s10772-015-9328-y
8. Rabiee A, Setayeshi S (2010) Persian accents identification using an adaptive neural network. In: 2nd international workshop on education technology and computer science. Wuhan, China, pp 7–10. https://doi.org/10.1109/etcs.2010.273
9. Rao SK, Koolagudi SG (2011) Identification of Hindi dialects and emotions using spectral and prosodic features of speech. Systemics, Cybern Informatics 9(4):24–33

10. Behravan H, Hautama¨ki V, Kinnunen T (2015) Factors affecting i-vector based foreign accent recognition: a case study in spoken Finnish. Speech Commun 66:118–129
11. Ma ZC, Fokoué E (2014) A comparison of classifiers in performing speaker accent recognition using MFCCs. Open J Stat 4:258–266. https://doi.org/10.4236/ojs.2014.44025
12. Biadsy F, Hirschberg J, Habash N (2009) Spoken Arabic dialect identification using phonotactic modeling. In: Proceedings of the EACL workshop on computational approaches to semitic languages. Athens, Greece, pp 53–61
13. Salau AO, Jain S (2019) Feature extraction: a survey of the types, techniques and applications. In: 5th IEEE international conference on signal processing and communication (ICSC-2019). Noida, India
14. Salau AO, Oluwafemi I, Faleye KF, Jain S (2019) Audio compression using a modified discrete cosine transform with temporal auditory masking. In: 5th IEEE international conference on signal processing and communication (ICSC-2019). Noida, India
15. Mukherjee R (2012) Speaker recognition using shifted MFCC. Graduate Theses and Dissertations, pp 1–56
16. Piat M, Fohr D, Illina I (2008) Foreign accent identification based on prosodic parameters. In: Interspeech. Brisbane, Australia, pp 759–762
17. Jiao Y, Tu M, Berisha V, Liss J (2016) Accent identification by combining deep neural networks and recurrent neural networks trained on long and short term features. In: Interspeech, pp 2388–2392. https://doi.org/10.21437/interspeech.2016-1148
18. Amuda SAY, Boril H, Sangwan A, Hansen JHL, Ibiyemi TS (2014) Engineering analysis and recognition of Nigerian English: an insight into low resource languages. Trans Mach Learning Artif Intell 2(4):115–126. https://doi.org/10.14738/tmlai.24.334
19. Chittaragi NB, Prakash A, Koolagudi SG (2018) Dialect identification using spectral and prosodic features on single and ensemble classifiers. Arabian J Sci Eng 43(8):4289–4302. https://doi.org/10.1007/s13369-017-2941-0
20. Faria A (2006) Accent classification for speech recognition. In: Renals S, Bengio S (eds) Machine learning for multimodal interaction, MLMI (2006). Lecture notes in computer science, 3869. Springer, Berlin, Heidelberg. https://doi.org/10.1007/11677482_25
21. Hammami N, Bedda M, Farah N (2012) Spoken Arabic digits recognition using MFCC based on GMM. In: IEEE conference on sustainable utilization and development in engineering and technology. Kuala Lumpur, Malaysia, pp 160–163. https://doi.org/10.1109/student.2012.6408392
22. Soorajkumar R, Girish GN, Ramteke PB, Joshi SS, Koolagudi SG (2017) Text-independent automatic accent identification system for Kannada Language. In: Satapathy S, Bhateja V, Joshi A (eds) Proceedings of the international conference on data engineering and communication technology. Advances in intelligent systems and computing. Springer, Singapore, p 469
23. Ullah S, Karray F (2007) Speaker accent classification system using fuzzy canonical correlation-based gaussian classifier. In: IEEE international conference on signal processing and communications. Dubai, pp 792–795. https://doi.org/10.1109/icspc.2007.4728438
24. Yusnita MA, Paulraj MP, Yaacob S, Yusuf R, Shahriman AB (2013) Analysis of accent-sensitive words in multi-resolution mel-frequency cepstral coefficients for classification of accents in Malaysian English. Int J Automot Mech Eng (IJAME) 7:1053–1073. https://doi.org/10.15282/ijame.7.2012.21.0086
25. Zheng Y, Sproat R, Gu L, Shafran I, Zhou H, Su Y, Jurafsky D, Starr R, Yoon S (2017) Accent detection and speech recognition for Shanghai-Accented Mandarin. In: Interspeech, pp 217–220

Chapter 2
Development of an Effective Computing Framework for Classification of Motor Imagery EEG Signals for Brain–Computer Interface

Pinisetty Sri Ramya, Kondabolu Yashasvi, Arshiya Anjum, Abhijit Bhattacharyya and Ram Bilas Pachori

1 Introduction

Brain–computer interface (BCI) is a developing technology whose intention is to pass on an individual's thought processes to the external world straightforwardly from their thinking process and thereby improving cognitive capacities. It is a direct communication channel between the mind and an outside assistive gadget [1]. The most commonly used method is the designing of a BCI system relying on the usage of electroencephalogram (EEG). An EEG signal is used to measure neurophysiological activity [2] in terms of electrical signals generated in the brain. Motor imagery [3] is a mental simulation of the task. Wolpaw et al. translated [1] the neuronal activities into real-time commands developing new augmentative communication and control technology which not only is concerned with the field of disability but also integrates different domains.

EEG-based motor imagery classification is coupled with certain difficulties such as noise addition [4] while recording, due to artifacts caused by muscular activity and poor electrode contact. EEG signals are subject specific, non-stationary, and are greatly prone to noise due to artefacts [5], which makes them difficult to analyse. Gaur et al. [6] proposed a novel pre-processing technique that automatically reconstructs the EEG signal by selecting the intrinsic mode functions (IMFs) based on median measure of frequency. The EEG signal which is reconstructed has shown high signal-to-noise ratio (SNR) and consists of only that information which is correlated with particular motor imagery task.

P. Sri Ramya (✉) · K. Yashasvi · A. Anjum · A. Bhattacharyya
National Institute of Technology, Tadepalligudem, Andhra Pradesh, India

R. B. Pachori
Indian Institute of Technology Indore, Indore, India
e-mail: pachori@iiti.ac.in

S. Jain et al. (eds.), *Advances in Computational Intelligence Techniques*, Algorithms for Intelligent Systems,
https://doi.org/10.1007/978-981-15-2620-6_2

All the traditional signal decomposing techniques like short-time Fourier transform [7], wavelet transform [8], etc., decompose a signal based on the assumption of linearity, leading to sub-optimal localization, resulting in inadequate performance. This has led to the development of a new data-driven adaptive technique known as empirical mode decomposition (EMD) [9], which has been established with acceptable performance in majority cases of nonlinear, multi-scale and real-time non-stationary data series. In order to cater the need of real-time multichannel series of data, Mandic et al. [10] developed an extension to EMD algorithm, known as multivariate EMD (MEMD).

A comparative study [5] which analyses the accuracy of MEMD in dealing with non-stationarity, closely spaced frequency bands and low SNR is made for BCI competition IV data set 1. This study recorded maximum accuracy of 91.5% for subject G. In [11], Suman Dutta used multivariate autoregressive models for six-channel EEG data in MEMD domain. Eigenvalues calculated from the co-variance matrices have shown a high distinguishing accuracy of 94.3%. Gaur [12] analysed the features that are extracted utilizing single-channel and multichannel EMD-based filtering for the classification of the multiple direction wrist movements based on magnetoencephalography (MEG) signals for improving BCI. This analysis proved that the multichannel filtering can be utilized in online BCI systems for effective results. Figure 1 shows the classification system for BCI application.

The objective of this work is to develop a composite method of adaptive signal processing and machine learning based on a new methodology for BCI motor

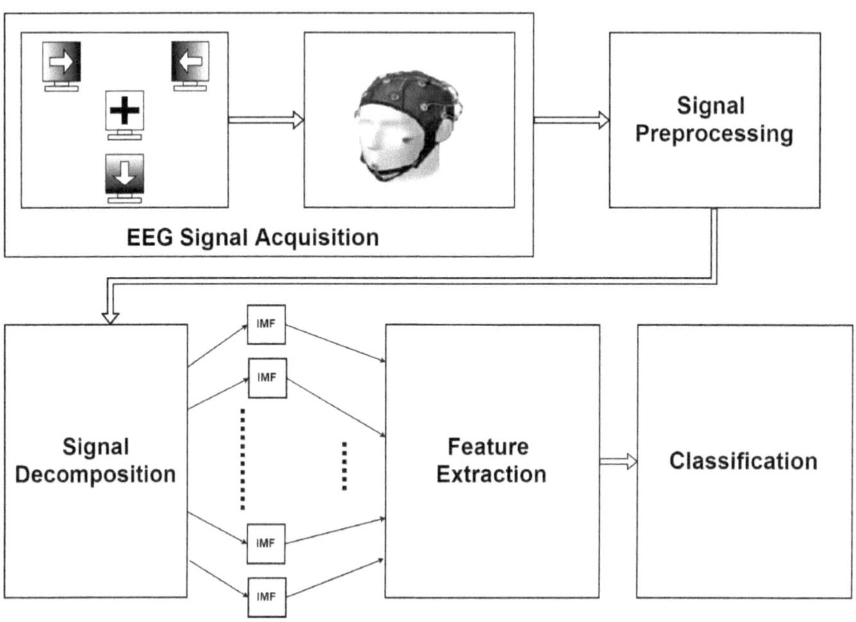

Fig. 1 Illustration of the experimental paradigm. Dotted lines show the time windows extracted from EEG signals

imagery classification using EEG signals which can be easily deployed in an effective computing framework, thereby providing disabled human beings with a new communique channel.

2 Empirical Mode Decomposition

The EMD [9] has been proposed as a data-adaptive method. In the EMD approach, data which is represented as $Y(t)$ is decomposed into number of IMFs [10]. IMFs are simple oscillatory mono-component functions with a varying frequency and amplitude. They are of the identical length as the source signal and conserve the frequency alterations with time. The signal can be reconstructed by simply summing up the IMFs as mentioned in (1) as follows:

$$Y(t) = \sum_{i=0}^{n} IMF_i + k_n \tag{1}$$

where k_n is the residue of the source signal $Y(t)$ and n is the number of extracted IMFs.

IMFs have the following properties:

1. In the entire data set, the total number of zero crossings and the number of extrema must be either differed by at most one or equal.
2. At any instant, the mean of the envelope formed by local minima and local maxima must be equal to zero.

2.1 The EMD Algorithm

In EMD implementation, sifting process which uses local extrema is applied. For the signal $Y(t)$, EMD algorithm is summarized as follows:

1. Detect all local extrema (a combination of maxima and minima) of the signal $Y(t)$.
2. Connect all these local maxima (minima) with corresponding splines based on the application, and generate the upper (lower) envelope $l_u(t)$ $(l_i(t))$.
3. Obtain the local mean $me(t)$ of the lower and upper envelopes using the following expression:

$$[me(t) = [l_u(t) + l_i(t)]/2]$$

4. Consider $de(t) = Y(t) - me(t)$ as the data, and reiterate steps 1 and 2 multiple times until the normalized squared difference between two consecutive sifting operations defined as in Eq. (2) to be minimal.

$$SD_i = \frac{\sum_{i=0}^{T} |\mathrm{d}e_{i-1}(t) - \mathrm{d}e_i(t)|^2}{\sum_{i=0}^{T} \mathrm{d}e_{i-1}(t)^2} \tag{2}$$

If this squared difference SD_i is lesser than the threshold value, the process will be terminated and can be treated as sufficient IMF and the equivalent residue $r(t)$ is computed by

$$r(t) = Y(t) - \mathrm{d}e_k(t) \tag{3}$$

3 Multivariate Empirical Mode Decomposition

Though it has been proved that EMD is very useful and data adaptive, it still leaves behind some aggravating difficulties unresolved. Few major shortcomings of EMD are mode mixing, non-uniqueness (different numbers of IMFs for different channels) and mode misalignment (absence of synchronization and alignment of frequency information among different modes). In addition to this, employing EMD to each and every channel of the multivariate signal is tedious and generates sub-optimal results due to the drawbacks stated. A multivariate extension of EMD is proposed for alignment of scale among different channels.

MEMD [13] is a comprehensive extension of EMD for signal analysis containing multiple channels, by directly functioning in multidimensional space of actual signal. In MEMD, the critical task of finding the local mean of a multivariate signal is done through the multiple real-valued projections of the actual signal along the direction vectors. The direction vectors of multidimensional space are chosen by using uniform angular coordinates and low-discrepancy point sets that arise from quasi-Monte Carlo methods.

The MEMD algorithm can be explained with the following steps:

1. Choose suitable direction vectors d^{a_l} which are samples (point set) on an $M-1$ sphere.
2. Obtain the projections of the signal $y(t)_{T=1}^{X}$ along the chosen multiple direction vectors for all L (where L is the total number of direction vectors), resulting in $pr^{\theta_l}(t)\}_{t=1}^{X}$ as projections.
3. Note the time points $x_K^{\theta}(t)$ that are correlated with the maxima of each of the projected signals.
4. Interpolate $x_L^{\theta}(t)$, $y(t)_{t=1}^{X}$ using a corresponding spline and generate multivariate envelope set $E^{\theta_L}(t)_{L=1}^{L}$.
5. For a total number of L direction vectors, the mean $me(t)$ of the envelope set is calculated as

$$me(t) = \frac{1}{L}\sum_{k=1}^{L} E^{\theta_k}(t)$$

6. Consider $de(t) = y(t) - me(t)$ as the new data, and if $d(t)$ fulfils the stopping criterion for an IMF, reiterate above procedure to $de(t) - y(t)$, else reiterate for $de(t)$.

4 Our Proposed Approach

4.1 *BCI IV Data Set 1*

The data [14] was taken from seven healthy subjects where three motor imagery actions are performed without feedback. The three motor imagery tasks are left, right and foot. For a specific subject, only two motor imagery actions were chosen and recordings were continuously taken by switching subject intentions randomly. Each subject performs a total of 200 trials with a 100 trials of each task as shown in Table 1 (Fig. 2).

In one trial, subject has imagined about one task and the length of one trial is about 4s. The data was taken from 59 EEG channels using a Ag/AgCl electrode cap and BrainAmp MR plus amplifiers. Out of these 59 channels, 11 channels were selected from the central part of the brain where the brain activity is maximum. They are 'FC3', 'FC4', 'Cz', 'C3', 'C4','C5', 'C6', 'T7', 'T8', 'CCP3' and 'CCP4' in accordance with the 10–20 systems. In this data set, the pre-processing of signals is done in two steps. They are segmentation based on task and filtration. The data of recordings was sampled at 1000 Hz initially and then downsampled to 100 Hz.

4.2 *Features*

The accuracy of the classification is decided by the features chosen to distinguish the signals [15]. The following features are used in this work.

Table 1 BCI competition IV data set 1: subjects and number of trials for respective tasks

Subject	Left task	Right task	Foot task
A	100	0	100
B	100	100	0
C	100	100	0
D	100	100	0
E	100	100	0
F	100	0	100
G	100	100	0

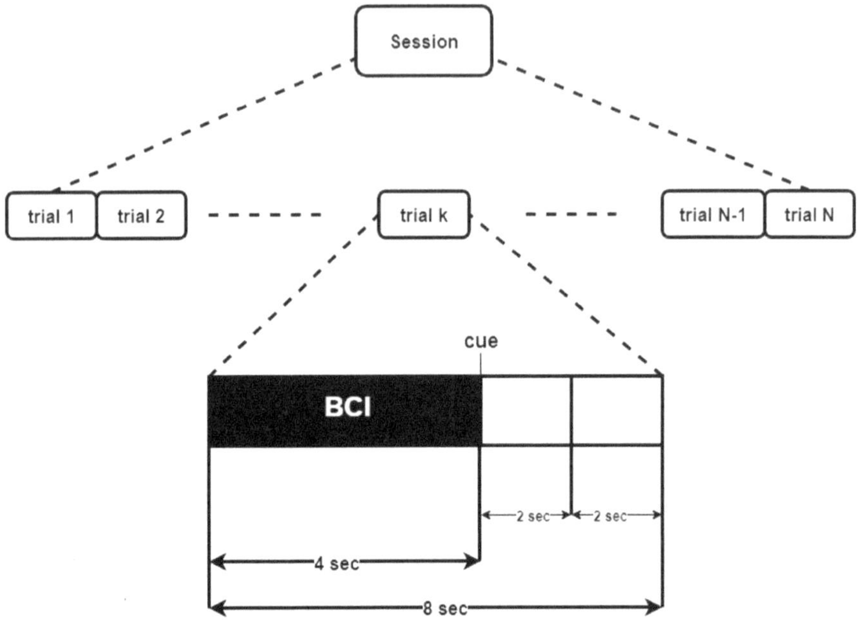

Fig. 2 Illustration of the experimental paradigm. Dotted lines show the time windows extracted from EEG signals

4.2.1 Skewness

Skewness is a measure of the proportion of asymmetry present in statistical distribution, where the curve seems misshaped either to the left or to the right. It can be used to describe to the degree to which a distribution contrasts from a normal distribution. It can be expressed mathematically as follows:

$$\text{skewness} = \sqrt{\frac{\sum_{j=1}^{n} \left(\bar{t} - t_j\right)^3}{(n-1)\, s^3}}$$

$\bar{t}$ Mean
s Standard deviation
n Sample size

4.2.2 kNN Entropy

kNN entropy depends on statistics of k-nearest neighbour (kNN) distances [16]. Let the probability density be $\phi_k(\varepsilon)$ of ε, the distance from x_i to its kNN. The density function $\phi_k(\varepsilon)\mathrm{d}\varepsilon$ indicates the probability of strictly one point which is in $[\varepsilon, \varepsilon + \mathrm{d}\varepsilon]$, strictly $k - 1$ points at a distance less than the kNN and the remaining points farther than the kNN. It is as follows:

$$\phi_k(\varepsilon)\mathrm{d}\varepsilon =$$

$$\binom{N-1}{1}\frac{\mathrm{d}\phi_i(\varepsilon)}{\mathrm{d}\varepsilon}\mathrm{d}\varepsilon\binom{N-2}{k-1}(\phi_i(\varepsilon))^{k-1}(1-\pi_i(\varepsilon))^{N-k-1}$$

The expected value of log(Pi) can be obtained as

$$E(\log\phi_i) = \int_0^\infty \log\phi_i(\varepsilon)\phi_k(\varepsilon)\mathrm{d}\varepsilon = \psi(k) - \psi(N)$$

$$\phi_i \approx \eta_i\phi_x(x_i)$$

$$\hat{H}(X) = \psi(N) - \psi(k) + \frac{1}{N}\sum \log\eta_i$$

4.3 Sample Entropy

It is utilized to compute the regularity in data without knowledge about a system. Initially, the vectors are constructed in a pseudo-phase: from a time series of N points $se(1), se(2), \ldots, se(N)$, k is the embedding dimension [17].

$$C_i^k(r) = \frac{(\text{number of } t(j) \text{such that } d[t(i), t(j)] \le r)}{N-k+1}$$

And r is a level of filtering, while d indicates distances of points. Generally, $r = 20\%$ of the standard deviation of the magnitude values and $k = 2$.

$$\phi^k(r) = \frac{1}{N-k}\sum_{i=1}^{N-k} C_i^k(r)$$

Similarly, $k + 1(r)$ can be defined as an embedding dimension of $k + 1$. The sample entropy is computed as:

$$\mathrm{Samp}_e n = \ln\frac{\phi^k(r)}{\phi^{k+1}(r)}$$

4.4 Fuzzy Entropy

It is almost similar to the related measure of sample entropy, but the concept of fuzzy sets is imported on the basis of exponential functions [18].

In calculating fuzzy entropy [17], initially the vectors constructed in the pseudo-phase are transformed as follows:

$$\left[\bar{t}(i)\right] = \left[se(i) - \overline{se}(i), \ldots, se(i + p - 1) - \overline{se}(i)\right] \quad (4)$$

$$\left[\bar{t}(j)\right] = \left[se(j) - \overline{se}(j), \ldots, se(j + p - 1) - \overline{se}(j)\right] \quad (5)$$

where $\overline{x}(i)$ is the mean value of t(i):

$$\overline{se}(i) = \sum_{k=0}^{p-1} \frac{se(i + k)}{p}$$

In the next step, the fuzzy membership matrix is defined as:

$$D_{i,j}^{p} = \mu(d(\overline{t_i}^{p}, \overline{t_j}^{p}))$$

with the fuzzy membership function:

$$\mu(se) = \mathrm{e}^{-\left(\frac{t}{r}\right)^{n}}$$

Finally, fuzzy entropy is defined as:

$$\text{Fuzzy}_{en} = \ln\frac{\phi^{p}}{\phi^{p+1}}$$

where

$$\phi^{p} = \frac{1}{N - p} \sum_{i=0}^{N-p} \sum_{j=1, j\neq i}^{N-p} \frac{D_{i,j}^{p}}{N - p - 1}$$

4.5 Permutation Entropy

The permutation entropy provides a method to quantify the complexity of time series [19]. Basically, each time series is associated with a probability distribution ϕ, whose elements ϕ_j are the frequencies connected with j feasible permutation patterns, where $i = 1, \ldots, n!$. Permutation entropy (P_{en}) is defined as

$$P_{en} = -\sum_{j=1}^{n!} \phi_j \ln \pi_j$$

Normalized permutation entropy is defined as:

$$P_{en_{\text{norm}}} = -\frac{1}{\log_2 n!}\sum_{j=1}^{n!} \phi_j \log_2 \phi_j$$

4.6 Classifiers

4.6.1 J48 Classifier

J48 is a decision tree algorithm which estimates the target variable class with the help of an ID3 tree classification approach [18]. Besides the simple and comprehensible distribution of data, J48 consists of the additional features such as decision tree pruning, accounting for missing values and the abstraction of rules. The aim is a continuous generalization until it reaches equilibrium in its accuracy as well as the flexibility of the tree.

4.6.2 Naive Bayes (NB) Classifier

Naive Bayes is one of the easy techniques for building classifiers based on a basic principle; i.e. it assumes that a specific feature is independent of other features when class attributes are given [20]. In spite of its amateur design and oversimplified hypothesis, this classifier has given satisfactory performance in case of complex real-world situations. The benefit of Naive Bayes is that this classifier requires a relatively less training set to determine the criterion which is essential for classification.

4.6.3 RF Classifier

RF classifier comprises a collection of structured decision tree classifiers $t(a, \theta_n)$, $n = 1, \ldots, k$ where θ_n are random vectors which are distributed independently. Every individual tree determines a poll for the common output class for given input x [21].

For an assemblage of trees $t_1(a), t_2(a), \ldots, t_n(a)$, and with the help of a training data set which is taken randomly from the random vector distribution R, a, the limiting function is detailed as

$$\text{Lmt}(A, R) = av_n \text{In}(t_n(A) = R) - \max_{j \neq R} av_n I(t_n(A) = j)$$

where In(.) is the index function.

The limit indicates the degree by which the mean number of votes at A, R for the right class surpasses the average decision in favour of some other tasks.

5 Results and Discussion

The performance of BCI depends on different evaluation parameters. Among these evaluation parameters of interest in this work is the classification accuracy. This section reports experimental evaluations of the classification accuracy.

The data set used in this study consists on average of 190,518 data points from 59 channels comprising all the 7 subjects. In this work, we have extracted four-second duration of EEG epochs from each of the 11 EEG channels for seven different subjects. This results in 30,800 numbers of left EEG epochs, 22,000 numbers of right EEG epochs and 8800 numbers of foot EEG epochs for motor imagery classes.

For all the IMFs, five features had been extracted which are kNN entropy, fuzzy entropy, sample entropy, permutation entropy and skewness. A correlation-based feature selection filter has been applied on the extracted features. Percentage split of 50% is used as criteria for the splitting of J48, RF and NB. The extracted IMFs using MEMD and EMD of the three motor imagery classes of EEG signals are shown in Figs. 3, 4, 5, 6, 7 and 8, respectively.

The IMFs obtained in the case of EMD are downsized to 5 minimum numbers of IMFs which are common to different epochs, whereas in MEMD the problem of non-uniqueness is eliminated and an identical number of 8 IMFs are obtained. It is evident from the IMFs of both EMD and MEMD that the rapid changes in the signal are captured first followed by the slow changes and each IMF is of the same size as that of the original EEG signal.

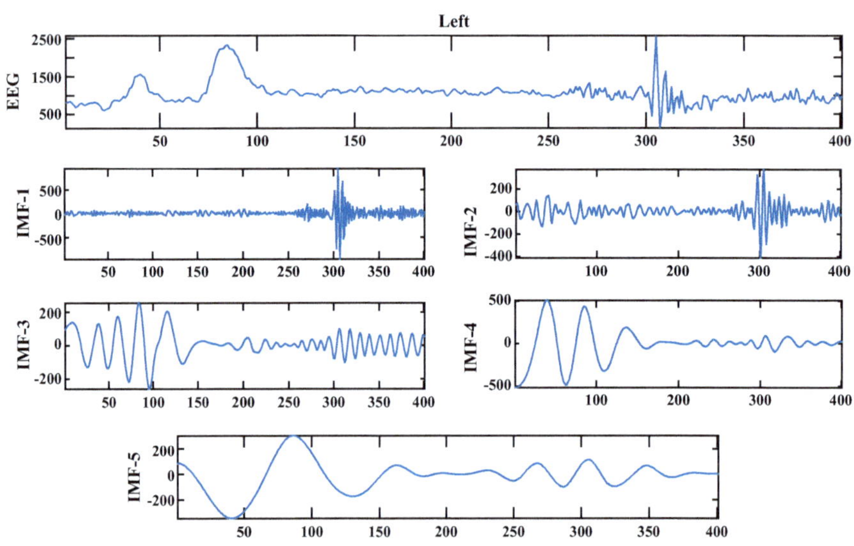

Fig. 3 Plot of EEG signal of FC3 channel and its extracted IMFs using EMD method for left motor imagery task of subject A

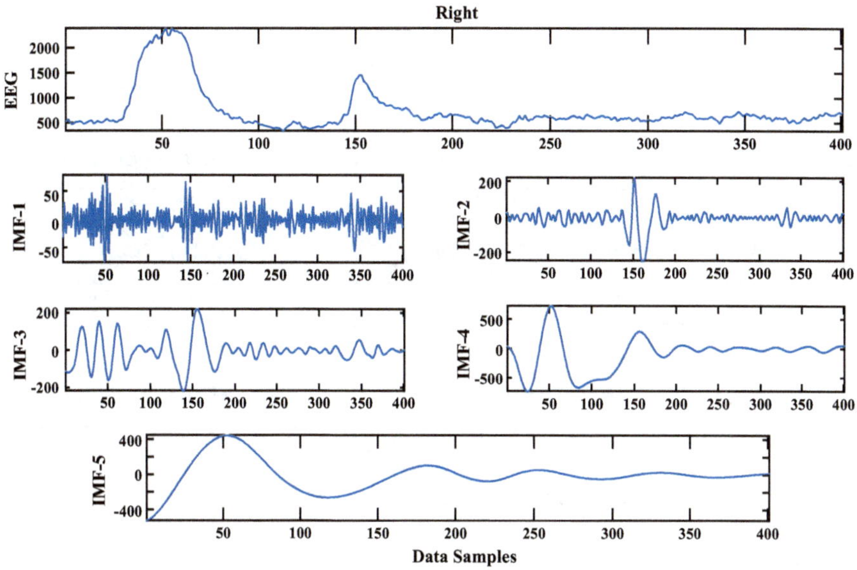

Fig. 4 Plot of EEG signal of FC3 channel and its extracted IMFs using EMD method for right motor imagery task of subject B

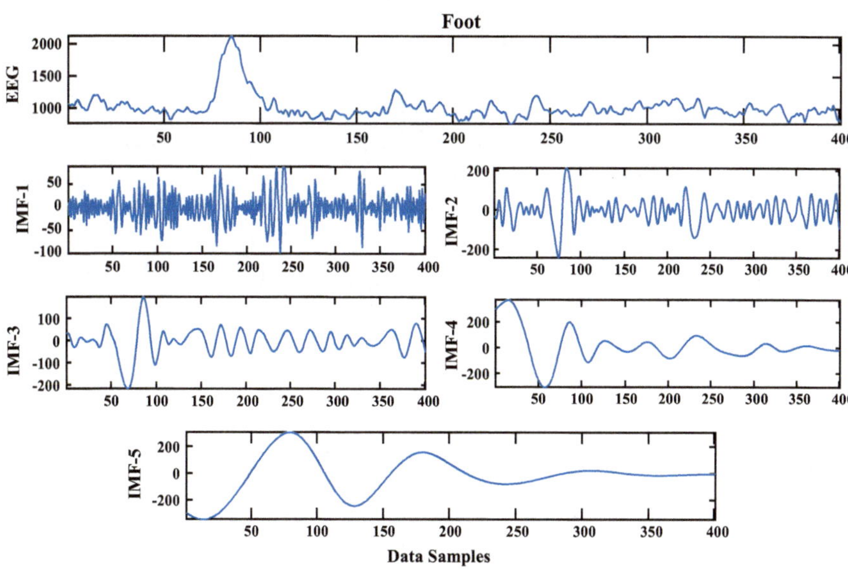

Fig. 5 Plot of EEG signal of FC3 channel and its extracted IMFs using EMD method for foot motor imagery task of subject A

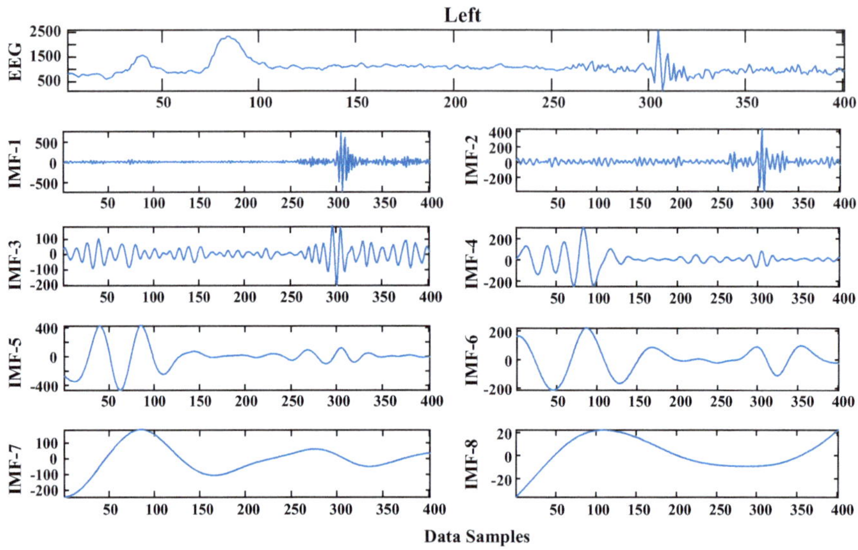

Fig. 6 Plot of EEG signal of FC3 channel and its extracted IMFs using MEMD method for left motor imagery task of subject A

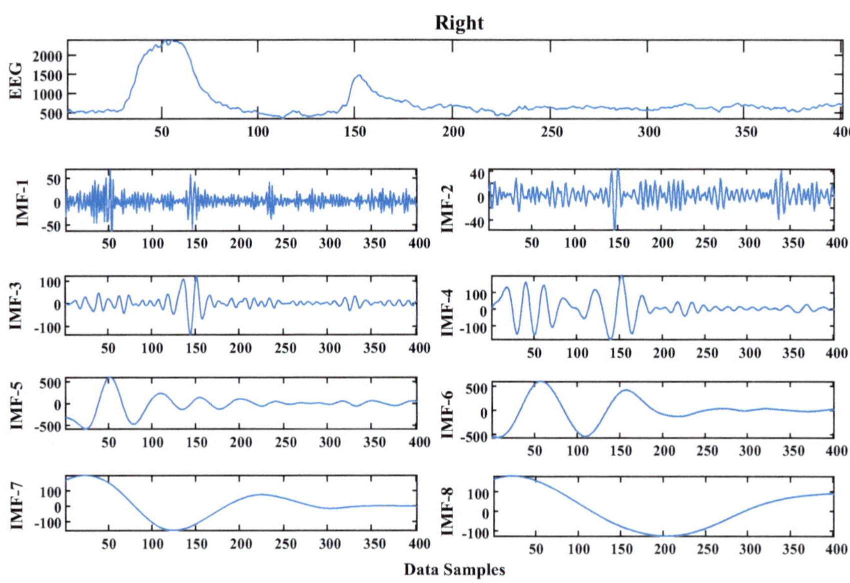

Fig. 7 Plot of EEG signal of FC3 channel and its extracted IMFs using MEMD method for right motor imagery task of subject B

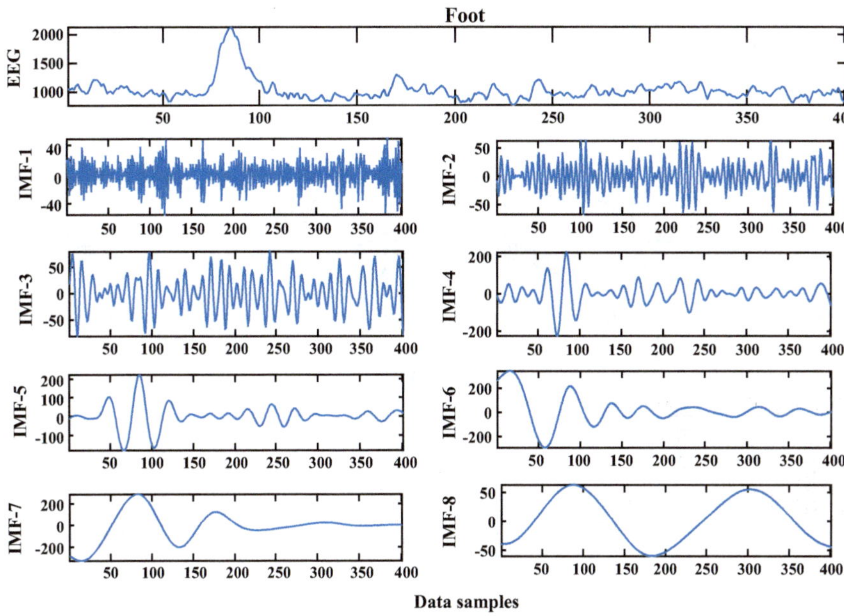

Fig. 8 Plot of EEG signal of FC3 channel and its extracted IMFs using MEMD method for foot motor imagery task of subject A

IMFs of a particular IMF group have common frequency of oscillation, but they originate from different cortical locations due to simultaneous decomposition of the six-channel EEG data by the MEMD algorithm alike channel-by-channel analysis by EMD.

After decomposition, feature extraction (kNN entropy, fuzzy entropy, sample entropy, permutation entropy, HFD and skewness) is performed for all the seven subjects. As mentioned earlier, 6 features are computed from each IMF resulting in 25 features per channel in case of EMD and 40 features per channel in case of MEMD. Therefore, the dimension of feature vector for EMD and MEMD is $\mathbb{R}^{25Cn \times 1}$ and $\mathbb{R}^{40Cn \times 1}$, respectively, where Cn indicates the number of channels.

The results in Figs. 9, 10 and 11 show the feature distribution box plots of IMF-1 for the channel FC3 corresponding to of first five best features, i.e. sample entropy, kNN entropy, fuzzy entropy, skewness and permutation entropy. It is evident from the figures that there is significant discrimination in the separability among three different classes for most of the features. Only five significant features are considered, although various other features are used for the training session and evaluated (Figs. 12 and 13).

Through the training of various features and our experimental observations, entropy feature has a predominant ability in classification of EEG signals than the rest of the time- and frequency-based features. Among the five, kNN entropy feature has the major impact in the result.

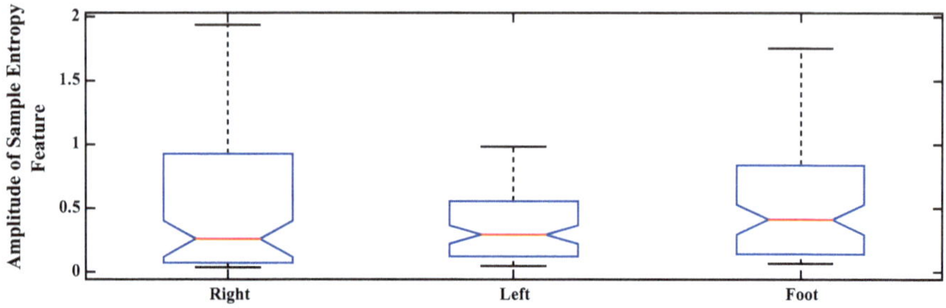

Fig. 9 Box plots obtained for right, left and foot motor imagery tasks using sample entropy feature for channel FC3 for IMF-1

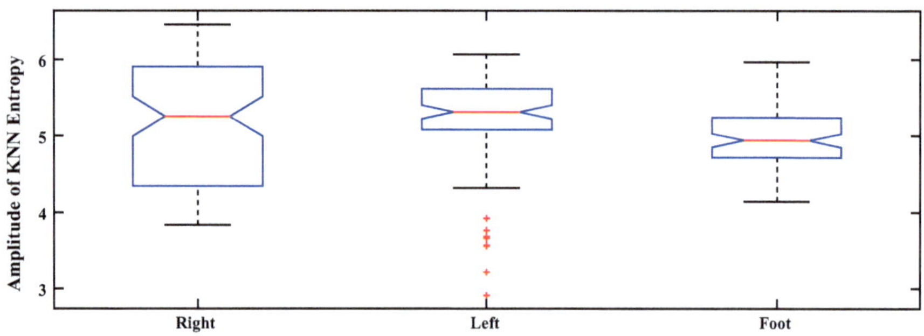

Fig. 10 Box plots obtained for right, left and foot motor imagery tasks using kNN entropy feature for the channel FC3 for IMF-1

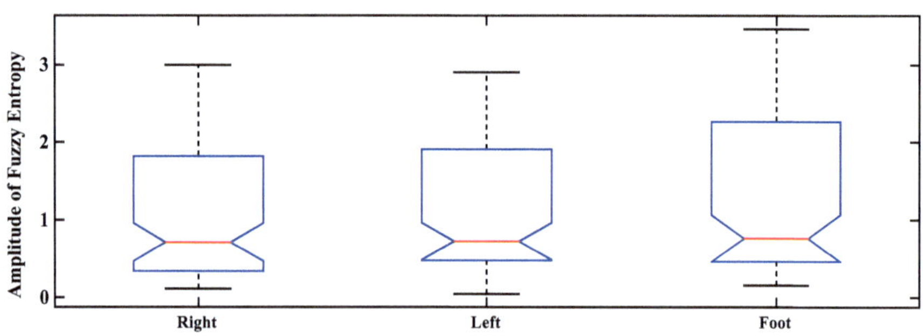

Fig. 11 Box plots obtained for right, left and foot motor imagery tasks using fuzzy entropy feature for channel FC3 for IMF-1

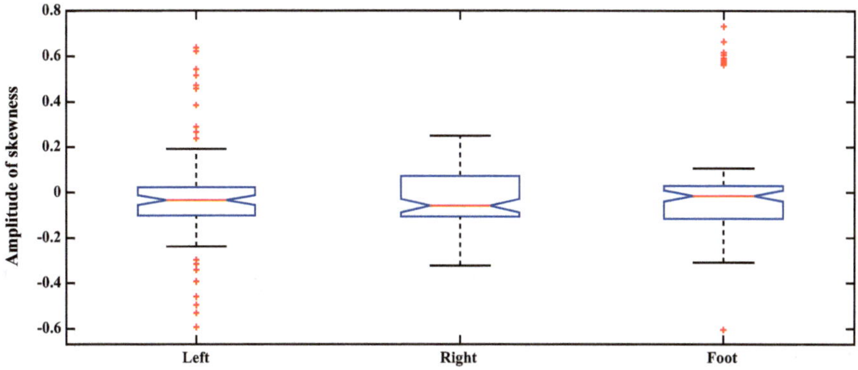

Fig. 12 Box plots obtained for right, left and foot motor imagery tasks using skewness feature for channel FC3 for IMF-1

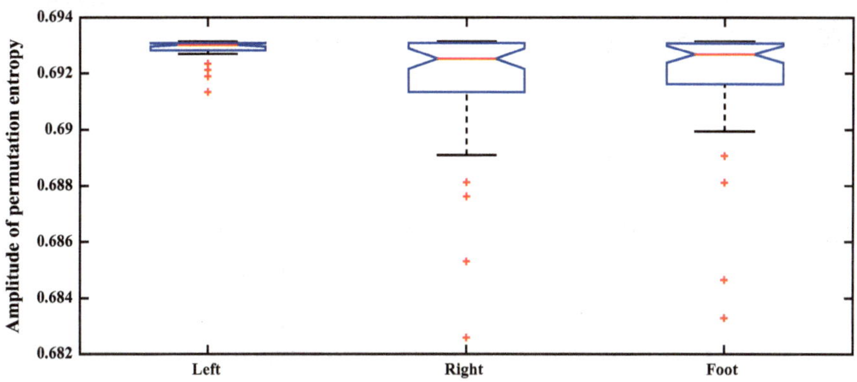

Fig. 13 Box plots obtained for right, left and foot motor imagery tasks using skewness feature for channel FC3 for IMF-1

In order to reduce the dimensions of feature vector, a filter called correlation-based feature selection (CFS) [22] is applied on the feature matrix. The processed features are then fed to classifiers (RF, NB and J48) using percentage spilt (50% of the feature vector as training and test data sets). A comparison of obtained classification accuracies for the extracted features using the three classifiers is presented in Table 2.

The detailed performance of MEMD for each individual subject is evaluated using tenfold cross-validation [23], and percentage split is presented in Tables 3, 4, 5, 6, 7, 8 and 9.

Tables 3, 4, 5, 6, 7, 8 and 9 also provide the details of the accuracies when a CSV filter is applied and when it is not. The impact of CSV filter is clearly seen, and improved results can be observed. (These tables are evaluated using the MEMD signal processing approach.)

There is a considerable increase in the classification performance using MEMD technique as compared to EMD technique. The maximum reported accuracies for

Table 2 Classification accuracy for J48, RF and NB classifiers for 11 channel data when MEMD and EMD algorithms are used

Subjects	J48 decision tree		RF		NB	
	EMD (%)	MEMD (%)	EMD (%)	MEMD (%)	EMD (%)	MEMD (%)
Subject A	54	57	55	58	54	56
Subject B	46	84.5	44	93	51	68.5
Subject C	49	89	52	98	51	94.5
Subject D	47	94.5	41	99.5	49	95.5
Subject E	54	92	55	99	46	81.5
Subject F	50	69	50	72	51	66.5
Subject G	49	64	55	74	52	65

Table 3 Obtained classification accuracies for J48, RF and Naive Bayes classifiers for subject A when MEMD algorithm is used

Classifiers	Cross-validation		Percentage split	
	With CFS (%)	Without CFS (%)	With CFS (%)	Without CFS (%)
J48	52	54	66.5	58
RF	52.5	55	52.5	59
NB	56	62	59.5	58

Table 4 Obtained classification accuracies for J48, RF and Naive Bayes classifiers for subject B when MEMD algorithm is used

Classifiers	Cross-validation		Percentage split	
	With CFS (%)	Without CFS (%)	With CFS (%)	Without CFS (%)
J48	84.5	86	86.5	87
RF	93	91	94	93
NB	68.5	59	92.5	93

Table 5 Obtained classification accuracies for J48, RF and Naive Bayes classifiers for subject C when MEMD algorithm is used

Classifiers	Cross-validation		Percentage split	
	With CFS (%)	Without CFS (%)	With CFS (%)	Without CFS (%)
J48	89	96	92	92
RF	98	98	98	96
NB	94.5	89	98.5	99

subjects B, C, D and E are obtained using MEMD technique along with RF classifier, and the accuracies are 93%, 98%, 99.5% and 99%, respectively. Similarly, in the case of remaining subjects, accuracies are improved using MEMD. Among the three classifiers, the reported average accuracies for all the seven subjects using J48, RF and Naive Bayes are 78.7%, 84.8% and 75.3%, respectively. From this, we can rank

Table 6 Obtained classification accuracies for J48, RF and Naive Bayes classifiers for subject D when MEMD algorithm is used

Classifiers	Cross-validation		Percentage split	
	With CFS (%)	Without CFS (%)	With CFS (%)	Without CFS (%)
J48	94.5	90	94.5	91
RF	99.5	99.5	99.5	99.5
NB	95.5	93	99.5	99.5

Table 7 Obtained classification accuracies for J48, RF and Naive Bayes classifiers for subject E when MEMD algorithm is used

Classifiers	Cross-validation		Percentage split	
	With CFS (%)	Without CFS (%)	With CFS (%)	Without CFS (%)
J48	92	85	92.5	94
RF	99	99	99	99.5
NB	81.5	78	99.5	99.5

Table 8 Obtained classification accuracies for J48, RF and Naive Bayes classifiers for subject F when MEMD algorithm is used

Classifiers	Cross-validation		Percentage split	
	With CFS (%)	Without CFS (%)	With CFS (%)	Without CFS (%)
J48	69	62	76.5	71.5
RF	72	71	79.5	77
NB	66.5	61	73.5	72

Table 9 Obtained classification accuracies for J48, RF and Naive Bayes classifiers for subject G when MEMD algorithm is used

Classifiers	Cross-validation		Percentage split	
	With CFS (%)	Without CFS (%)	With CFS (%)	Without CFS (%)
J48	64	67	71.5	65
RF	74	70	77	74
NB	65	59	68.5	59

the efficiency of classification of the three classifiers used in our study with RF having the best accuracy followed by J48 and Naive Bayes.

The average accuracies across all the subjects using percentage split (88.78%) are comparatively higher than the same reported for tenfold cross-validation (83.65%). It is evident from the tables that the application of CFS filter for most of the cases has provided improved classification accuracy. This resulted into average classification accuracies of 87.3 and 85.3% for with and without the application of CFS filter.

Thus, it can be stated that the application of MEMD and CFS filter has enhanced the overall performance of the proposed technique for most of the considered subjects.

By this have arrived at a computational framework for classification, which is a combination of selecting the proper channels for the analysis, adaptive signal processing techniques, efficient feature set, feature selection, classifier and classification way.

6 Conclusions

A good performance accuracy has always been a challenge in BCI motor imagery domain. In this study, we have developed a combination of adaptive signal processing inspired from the empirical methods (EMD and MEMD) along with a proper feature set (kNN entropy, fuzzy entropy, sample entropy, permutation entropy and skewness) which yielded an optimal performance for 4 out of 7 subjects. This method outperforms the existing work carried out on this database.

In future extension of work, it may intrigue to assess the proposed technique for classification issue in real-time BCI and help the people. Furthermore, to enhance the proposed technique to characterize the EEG signal to classify multiple (more than 2 at a time) tasks. In addition to this, a conventional algorithm which can lower the number of EEG channels without trading off the classification accuracy can also be developed in future.

References

1. Wolpaw JR, Birbaumer N, McFarland DJ, Pfurtscheller G, Vaughan TM (2002) Brain-computer interfaces for communication and control. Clin Neurophysiol 113(6):767–791
2. Jahankhani P, Kodogiannis V, Revett K (2006) EEG signal classification using wavelet feature extraction and neural networks. In: IEEE John Vincent Atanasoff 2006 international symposium on modern computing (JVA'06)
3. Qin L, Ding L, He B (2004) Motor imagery classification by means of source analysis for brain-computer interface applications. J Neural Eng 1(3):135
4. Pijn JP, Van Neerven J, Noest A, da Silva FHL (1991) Chaos or noise in EEG signals; dependence on state and brain site. Electroencephalogr Clin Neurophysiol 79(5):371–381
5. Park C, Looney D, urRehman N, Ahrabian A, Mandic DP (2012) Classification of motor imagery BCI using multivariate empirical mode decomposition. IEEE Trans Neural Syst Rehabil Eng 21(1):10–22
6. Gaur P, Pachori RB, Wang H, Prasad G (2019) An automatic subject specific intrinsic mode function selection for enhancing two-class EEG based motor imagery-brain computer interface. IEEE Sens J
7. Allen J (1977) Short term spectral analysis, synthesis, and modification by discrete fourier transform. IEEE Trans Acoust, Speech, Signal Process 25(3):235–238
8. Daubechies I (1990) The wavelet transform, time-frequency localization and signal analysis. IEEE Trans Inf Theory 36(5):961–1005
9. Huang NE, Shen Z, Long SR, Wu MC, Shih HH, Zheng Q, Yen NC, Tung CC, Liu HH (1998) The empirical mode decomposition and the hilbert spectrum for nonlinear and non-stationary time series analysis. Proc R Soc London Ser A Math Phys Eng Sci 454(1971):903–995

10. Mandic DP, ur Rehman N, Wu Z, Huang NE (2013) Empirical mode decomposition-based time-frequency analysis of multivariate signals: the power of adaptive data analysis. IEEE Signal Process Mag 30(6):74–86
11. Dutta S, Singh M, Kumar A (2018) Automated classification of non-motor mental task in electroencephalogram based brain-computer interface using multivariate autoregressive model in the intrinsic mode function domain. Biomed Signal Process Control 43:174–182
12. Gaur P, Kaushik G, Pachori RB, Wang H, Prasad G (2019) Comparison analysis: single and multichannel EMD-based filtering with application to BCI. In: Machine intelligence and signal analysis. Springer, pp 107–118
13. Rehman N, Mandic DP (2009) Multivariate empirical mode decomposition. Proc R Soc A Math, Phys Eng Sci 466(2117):1291–1302
14. Zhang H, Guan C, Ang KK, Chin ZY (2012) BCI competition IV-data set I: learning discriminative patterns for self-paced EEG-based motor imagery detection. Front Neurosci 6:7
15. Gupta A, Agrawal R (2012) Relevant feature selection from EEG signal for mental task classification. In: Pacific-Asia conference on knowledge discovery and data mining. Springer, pp 431–442
16. Kraskov A, Stögbauer H, Grassberger P (2004) Estimating mutual information. Phys Rev E 69(6):066138
17. Azami H, Fernández A, Escudero J (2017) Refined multiscale fuzzy entropy based on standard deviation for biomedical signal analysis. Med Biol Eng Comput 55(11):2037–2052
18. Khushaba RN, Al-Jumaily A, Al-Ani A (2007) Novel feature extraction method based on fuzzy entropy and wavelet packet transform for myoelectric control. In: 2007 international symposium on communications and information technologies. IEEE, pp 352–357
19. Bandt C, Pompe B (2002) Permutation entropy: a natural complexity measure for time series. Phys Rev Lett 88(17):174102
20. Lewis DD (1998) Naive (bayes) at forty: the independence assumption in information retrieval. In: European conference on machine learning. Springer, pp 4–15
21. Pal M (2005) Random forest classifier for remote sensing classification. Int J Remote Sens 26(1):217–222
22. Hall MA (2000) Correlation-based feature selection of discrete and numeric class machine learning
23. Kohavi R et al (1995) A study of cross-validation and bootstrap for accuracy estimation and model selection. In: Ijcai, vol 14. Montreal, Canada, pp 1137–1145

Chapter 3
Automatic Glaucoma Diagnosis in Digital Fundus Images Using Deep CNNs

Ambika Sharma, Monika Agrawal, Sumantra Dutta Roy and Vivek Gupta

1 Introduction

Globally, eye diseases are increasing at an alarming rate. The top three eye diseases are cataract, glaucoma, and age-related macular degeneration (AMD). Talking about "glaucoma", which is the second leading cause of blindness in the world and number one blinding disease among African Americans, scientific evidences show that if early diagnosis and treatment are given during early stages, then permanent loss of vision can be prevented in at least 60% of the cases [1]. Thus, there is a pressing need to develop a cost-effective computer-based automatic diagnostic system that can assist medical experts in early diagnosis, thus save their time and efforts wasted on the analysis of healthy people. Glaucoma is a chronic eye disease in which the optic nerve gets gradually damaged and thus leads to permanent loss of vision [2, 3]. It is also called the "silent thief of sight" because the loss of sight usually occurs over a long period of time. Eye experts use specialized devices for the eye monitoring such as a slit lamp or an ophthalmoscope to look at the back of the eye. These ophthalmologists are able to evaluate the health of an eye by looking at the various

A. Sharma (✉)
Bharti School of Telecommunication Technology and Management (BSTTM), IITD, New Delhi, India
e-mail: Ambika.Sharma@dbst.iitd.ac.in

M. Agrawal
Centre for Applied, New Delhi, India
e-mail: Monika.Aggarwal@care.iitd.ac.in

S. D. Roy
Electrical Department, IITD, New Delhi, India
e-mail: sumantra@ee.iitd.ac.in

V. Gupta
AIIMS Delhi, New Delhi, India
e-mail: vgupta@aiims.edu

S. Jain et al. (eds.), *Advances in Computational Intelligence Techniques*, Algorithms for Intelligent Systems,
https://doi.org/10.1007/978-981-15-2620-6_3

characteristics of cup and disc such as colour, contour, and diameter and thus able to predict the presence of disease based on their experiences and knowledge. Usually, for the glaucoma diagnosis, the two main indicators are the cup-to-disc ratio (CDR) and inferior-superior-nasal-temporal (ISNT) area rule [4], and for the calculation of these parameters, it is necessary to segment the optic disc and cup simultaneously. Figures 1, 2, and 3 show some of the examples of healthy and glaucomatous retinal images from the database. Figure 2 clearly shows the glaucoma image with high cup-to-disc ratio, and Fig. 3 shows the inferior rim loss retinal image, respectively. As mentioned previously, glaucoma is frequently known as "the sneak thief of sight" because of gradual rise in intraocular pressure and damage of the optic nerve or vision. The disease shows no early symptoms in the majority of the cases [5]. Thus, it is important to raise awareness among mass public for early detection and regular examination of eyes as this is still successfully controllable at the early stage if treated properly.

A number of algorithms have already been proposed for the glaucoma detection [5–21]. Researchers have categorized the detection into two parts. In the first category, the diagnosis requires the complete optic disc and cup boundary knowledge for the cup-to-disc feature calculation and other experimentation's [7, 11, 13, 14, 22], whereas the other approach deals directly with the complete image and learns its own set of feature for final decision [15, 21]. The first approach includes various methods, and some of them are as follows. Wong et al. [23], citeb26 suggested level set method for disc extraction and a blood vessel kink detection along with variational level set-based method for cup segmentation. An ellipse is fitted to the obtained cup boundary so as to get an initial estimate based on pallor and to the disc boundary to get a regular shape of region of interests. The method used canny edge

Fig. 1 Healthy image

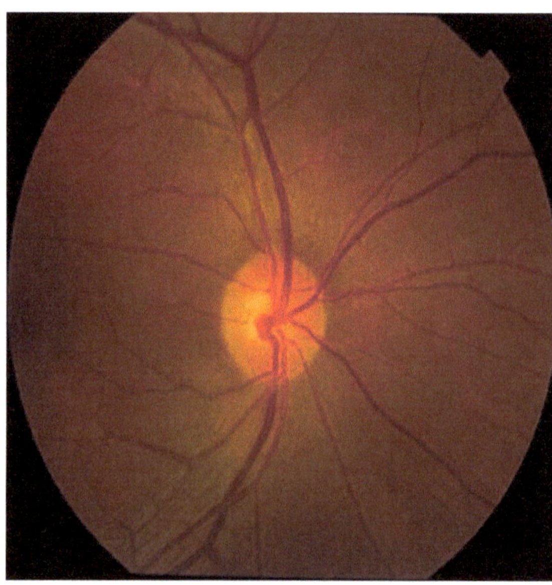

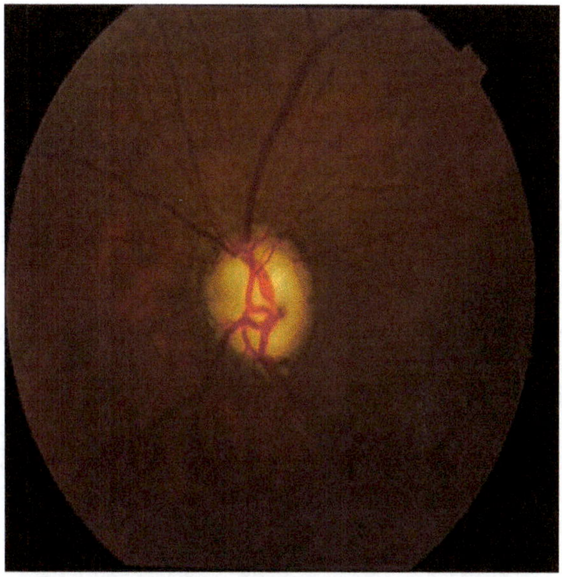

Fig. 2 High CDR (glaucoma)

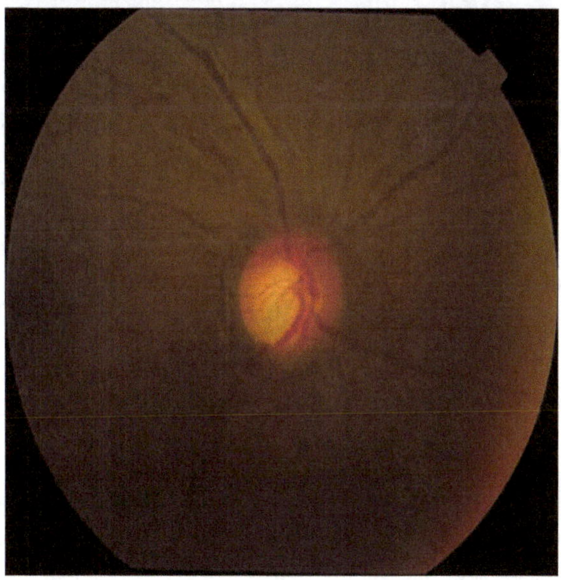

Fig. 3 Inferior rim loss (glaucoma)

detection along with the wavelet transform on the green channel image to detect the kinks on the intra-optic disc vessel edges. The extended work of this is mentioned in [24] which uses segmented disc and cup regions from level set approach to find the cup-to-disc Ratio (CDR) for glaucoma analysis. Cheng et al. [25] has also suggested an efficient approach based on super-pixel classification. The algorithm uses a simple linear clustering algorithm to collect the nearby pixels into superpixels. A set of OD and OC features have been constructed using histogram equalization of all three components. A support vector machine (SVM)-based library was used to classify the pixels as disc or non-disc in the OD and for OC segmentation, respectively. Yin et al. [26] have introduced model-based method along with circular Hough transform (CHT) for detection. From the last few years tremendous amount of effort has been done in glaucoma detection using machine learning techniques. Support vector machine, K-nearest neighbour, regression models, neural network, fuzzy min–max neural network, decision tree-based algorithms are some of the key techniques to perform automatic detection of glaucoma. The work proposed in [27] uses CNNs to learn the hierarchical information of retinal image and performs the optic disc and cup segmentation. The approach uses two neural networks, one for disc segmentation followed by a deep neural network of cup segmentation. For glaucoma prediction, it uses radii ratio of cup over disc along with the square root of areas of cup over disc. Recently, a novel approach proposed by Chen [21] where the network has used four convolutional layers and two fully connected layers. The network inputs the cropped OD regions for training and testing of the network. The obtained AUC values are 0.831 and 0.887 for ORIGA and SCES datasets. Another novel work by [28] utilizes sliding window idea using deep learning for glaucoma diagnosis. The system uses a bundle of sliding windows of different sizes to obtain the cup regions in each disc image, cropped from complete retinal image. It then extracts the features corresponding to each cup candidate using histogram method learned using a group sparsity constraint. The approach uses support vector regression model to rank each candidate region, and final prediction is made using non-maximal suppression method. In glaucoma screening, cup extraction has been suggested to be the toughest task. Thus, [29] suggested an efficient reconstruction approach to localize cup region for glaucoma diagnosis. The method has a code-book of reference disc and cup images, and the problem of finding the coordinates and radius of cup has been modelled as an optimization problem with the sparsity constraint. The results have been validated on SCES and ORIGA dataset. One of the key contributions in glaucoma screening is provided by [30] which uses U-Net for optic disc and cup segmentation and the uses four ImageNet-pre-trained CNN networks to classify glaucoma disease. Further, [31] uses cup-to-disc ratio as the key indicator for glaucoma and thus utilizes segmentation algorithms to find the disc and cup boundaries. The paper starts with OD detection method using the template matching and brightness property of optic disc. Later, it uses this apriori knowledge of optic disc area to find the disc and cup boundaries using the texture-based and model-based approach. The method achieves 98% accuracy on final glaucoma screening and diagnosis for Drishti dataset. Mvoulana et al. [31] work has been based on deep convolutional neural networks to segment the optic disc and cup regions for glaucoma diagnosis. The network is motivated by the DenseNet architecture which uses U-shape network to perform the pixel-wise classification. Another important work in the field is proposed by [32], where the paper has done an extensive study of different convolutional network for glaucoma diagnosis purpose. The method has used all existing classification networks to validate the retinal images as healthy or glaucoma. It has also incorporated the pre-trained weights of Imagenet dataset on existing architectures. Another approach has been suggested by [33] to diagnose glaucoma using deep convolutional neural network (CNN) unsupervised architecture which extracts the multilayer features from raw pixel intensities. Afterwards, the deep-belief network (DBN) model was used to select the most discriminative deep features based on the training dataset. At last, [34, 35] presents other work on glaucoma diagnosis using deep convolutional neural networks. Motivated by recent success in deep learning in biomedical image disease diagnosis, we have proposed a deep CNN architecture. As discussed, earlier CNN has wider applications in various domains right from face detection by Facebook, photo search by Google and recommendation models by Amazon.

The performance of all non-deep learning techniques depends upon the features being selected for classification. Also, in case of pathological conditions (presence of lesions, exudates, and light artifacts), it becomes difficult to precisely extract the suitable features for classification. Thus, it is important for a model to learn all the best possible features to extract, maximum possible information from the image, even in the presence of some other artifacts say light, or lens aberrations, thus avoids the need of hand-crafted features. The key idea behind our approach is to design an algorithm which incorporates both set of features, i.e. local and global, and enables us to predict the results with better accuracy. The obtained detection accuracy is 99.98% on the training and on tested dataset is 90.5%. The proposed algorithm has been discussed in following sections. Section 2 gives the brief introduction and motivation for proposed algorithm. In Sect. 3, some light has been shed on understanding the concept of deep neural networks. Later, Sect. 4 briefly introduces the proposed deep architecture for glaucoma diagnosis. At last, Sect. 5 shows the experimental and simulated results, followed by the conclusion in Sect. 6 for the proposed work.

2 Proposed Algorithm

During an eye examination, ophthalmologist uses retinal (fundus) images to examine the health of an eye. Retinal images are 2-D projection of 3-D retinal semi-transparent tissue (inner surface of eye) on an image plane. It consists of macula and optic disc along with vessels converging at it [4]. Optic disc is basically a bright yellowish and circular region in a healthy retinal image, but in case of pathological conditions, the appearance, shape, and colour of the optic disc might change [5]. Figure 4 shows the complete retinal/fundus image with all psychological features (optic disc, optic cup, macula, and blood vessels).

In case of glaucoma disease, the optic nerve gets gradually damaged, and one of the key factors for this damage is high intraocular pressure (IOP) inside the eye [4]. In a healthy eye, pressure varies from 10 to 21 mm Hg, but for the case of ocular hypertension, this value might go above 21 mm of Hg. While clinical examination ophthalmologists perform various diagnostic tests such as tonometry (measure

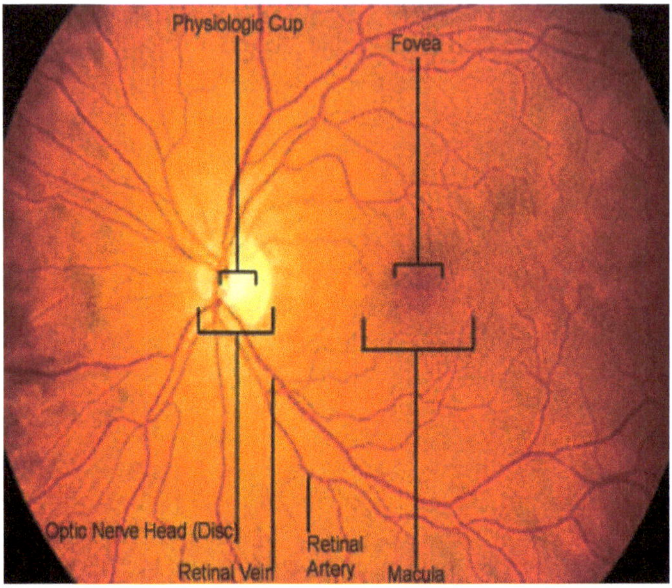

Fig. 4 Normal psychological parts of fundus image

intraocular pressure), pachymetry (determine cornea thickness), gonioscopy (measure drainage angle), disc photography, and visual field test/perimetry (gold standard test) to well diagnose the disease. All these methods require high end technology which is expensive and need well-skilled eye experts to analyse the results. Out of these tests, retinal image analysis is so far the most reliable approach to analyse the disease. The most important indicators for glaucoma are cup-to-disc (CDR) and inferior-superior to nasal-temporal (ISNT) ratios, which can be easily visualized in retinal image photography. The ratio of the diameter of optic cup-to-disc is expressed as a cup-to-disc ratio (CDR). There exists a wide range of C-D ratios for a normal eye [7–9]. In spite of being a strong clinical feature for glaucoma diagnosis, its implementation is really tedious, as finding the cup, and disc boundary is time consuming and requires skilled opticians to annotate the images. Thus, regular/or periodic monitoring of fundus photographs of the optic nerve is required to detect the gradual changes. This point forces one to think to have a device or digital software system which automatically diagnose the eye and give best possible results. Till data researchers have used the traditional approaches of image processing to diagnose the condition of eye. But, with the advent of big data and new technologies, deep learning approaches have become really popular for classification, segmentation, recognition, and various other applications. The deep learning architectures have the ability and flexibility to learn the nested hierarchical features used to represent the world. Its ability to learn generic features like edges, corners etc at bottom level, makes it really interesting to work with. This idea of learning hidden patterns in an image by CNN really motivates us to study and analyse the performance of various architectures with modified layers. In this paper, an efficient and accurate CNN architecture has been proposed to diagnose glaucoma in retinal images. The selected deep learning network consists of six layers, four consists of convolutional, and last two are fully connected classification layers. The network learns the feature representation of images from large dataset in a supervised end to end fashion.

3 Outline for Deep Learning Convolutional Neural Network

In the fields of self-driving cars, healthcare, finance, robots, and more, the deep learning (DL) has contributed a lot and changed the entire picture in terms of output efficiency. Under the architectures of DL, the convolutional neural networks (CNNs) have been applied broadly to the images and videos for various applications, and one of them is biomedical imaging. The CNN architectures use the nonlinear mapping and spatial scalability to progressively extract the high-level features from the raw input [17]. The hierarchical pattern present in the data is the basic intuition behind the implementation of these networks. For reference, in image processing or computer vision, the lower layers learn the abstract details about the image like edges, blobs, etc., whereas the higher layers recognize the fine and more meaningful details of the input such as alphabets/digits or faces [18].

For the proposed work, we require the CNN architecture as a classification network. In general, the CNN architectures mainly consist of four major components, which are convolutional layer, pooling layer, activation functions, and fully connected layers. We have briefly described each of these as follows.

3.1 Convolutional Layers

These are the basic building blocks in any CNN network, and the raw input image is fed to this layer of network. It simply performs the convolutional operation between the raw image and the designed filter, thus gives a feature map for different filters of varying sizes. The convolutional layer mainly learns the basic patterns present in the image, i.e. edges, blobs and contours, etc.

Mathematically, let $L^{(n-1)}$ and $L^{(n)}$ to be the input and output of nth layer, respectively. Also, it consists of filters of small spatial dimension, which slides over the entire input and gives the feature maps after giving the dot product between filter and image at each location. The filter values are learnable and updated after each new image set. For a CNN network with M layers, L^0 implies the input raw image, and L^M denotes the final output map of nth layer. Finally, we can represent the output of each convolutional layer as a linear mapping of inputs, i.e.

$$L^n = L^{n-1} * w^n + b^n, \quad (1)$$

where w^n, b^n represents the weights and biases for the nth layer. The output dimension of each layer can further be calculated using $(I–F + 2P)/S + 1$, where I is input image size, F is filter size, P is amount of zero padding done, S is stride applied. This layer extracts the deep hierarchical features present in the image, which in general, cannot be seen with naked eye.

3.2 Pooling Layer

A deep CNN networks have millions to billions of trainable parameters. Thus, to reduce the resources required for training such large number of parameters, a pooling layer has been introduced which reduces the spatial dimension of an image by a specified ratio. The most common filter size is 2×2 with stride 2 is applied to reduce the image dimension by half. In general, pooling is categorized into average and maximum pooling, where in the former case, average of four numbers (within 2×2 region) in latter maximum is computed, respectively. In general, pooling layers sum up the statistical features of a feature over the local regions. Apart from all of this, pooling also helps to maintain the translation invariant property in the image. Any local shift/translation at the local pixel level can be made invariant with the pooling operation [36], thus summarizes the output at local neighbourhood.

3.3 Activation Function

CNN tries to learn the nonlinear mapping present in the input image in order to extract the useful hidden features present and solve complex tasks. This function of nonlinearity is performed by the activation functions. It generally follows the convolutional layers in the CNN architectures. As previously mentioned, the convolutional output of each layer by Eq. (1) can be further modified as

$$L^n = f(L^{n-1} * w^n + b^n),$$

where f is the activation function used. In general, the most common activation functions are sigmoid or logistic, tanh and Relu. These functions are used for all convolutional layers except the last final convolutional layer.

3.4 *Fully Connected Layer*

It is the final fully connected convolutional layers and uses softmax as the activation function. These layers are more specific to the datasets and learn more abstract and detailed information of the data. The various tasks performed by CNN network such as classification, segmentation, localization are decided by the pattern of these layers. These layers represent the feature vectors which hold the aggregated an composite information from all previous convolutional layers.

4 Proposed Deep Architecture for Glaucoma Classification

The proposed convolutional neural network architecture uses four convolutional layers with two fully connected layers, will extract the hierarchical information of images to classify between glaucoma and non-glaucoma images. The details about the network have been explained in the following subsection and the complete architecture has been shown in Fig. 5.

4.1 *Region of Interest (ROI) Extraction and Data Augmentation*

Unlike the Chen [21] implementation, the network inputs the cropped optic disc as the region of interest of glaucoma detection, and we have used complete retinal image as the region of interest for the problem in hand. The reason is to encounter all possible features present in the complete image, including the key indicators outside the optic disc boundary such as retinal nerve fibre layer (RNFL) defect which suggested to be a strong measure for glaucoma after rim loss.

In the first step of extraction, the image has been cropped such that only field of view (FOV) is present, which avoid the unnecessarily information present in the background and also reduce the noise presence. In the latter step, the images have

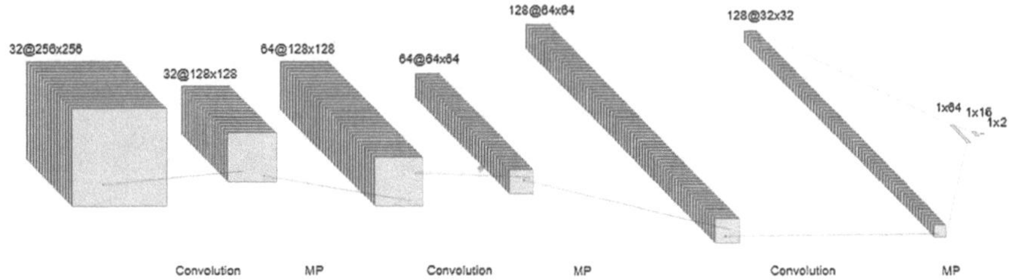

Fig. 5 Proposed architecture

been downscaled to 256 × 256 to reduce the time complexity of algorithm. In deep learning, the models do not generalize well on small datasets, thus suffer from overfitting problem. Although, in practice, the amount of data is limited. Some of the techniques to avoid this are to add normalization term, dropout layers, data augmentation or synthetically generate new images using methods like synthetic minority oversampling technique (SMOTE), generative adversarial networks (GANs) [20], etc. In object recognition problems, data augmentation has been observed as an effective technique [36]. In our proposed work, we employed data augmentation technique to increase the data points/samples, which enhance the variety of data for training. The various transformations used for data augmentation are horizontal flipping, vertical flipping, rotations with an increment of 30° within range of 0° to 360°, horizontal and vertical shifts within 0.2 range, and zooming in/out with 0.3, centre normalization and contrast enhancement. A combination of some of these transformations has also been employed to increase the effective accuracy of the proposed model.

4.2 Layer Architecture and Regularization Methods

The glaucoma classification problem has been modelled as the binary classification problem, where the input is the RGB retinal image and output is a scalar value, predicting the probability of the input image being a glaucomatous image. The architecture consists of four convolution layers with two fully connected layers. The best network has been selected after experimenting with various number of convolutional layers, fully connected neurons count, and different filter sizes. We have used the rectified linear unit (ReLu) activation function for all the convolutional layers except the last. This function is nonlinear in nature and less computationally expensive than "tanh" and "sigmoid" as it involves simpler mathematical operations. Also, it deals with the vanishing gradient problem where in case of low or zero error, the gradient of error is not able to reach the starting layers. For training the network, we have used binary cross entropy loss function to optimize the objective, and ADAM optimizer with learning rate of 0.00001 has been employed to achieve the minima for the cost function. The network has been trained for 200 epochs, with a batch size of 16. The final output layer uses the sigmoid activation for classifying the input retinal images. While training the network, binary cross entropy loss function is computed, and the system parameters (weights) are updated after each epoch. During testing, the image is resized and fed to the trained model for classification. The complete architecture, specifying the size of each filter, along with the dimension has been shown in Table 1.

Table 1 Detailed pipeline for the proposed architecture

Layer name	Output size	Filter size	Kernels	Stride
Conv1	256 × 256	3 × 3	32	1
MAX Pool	128 × 128	2 × 2	32	2
Conv2	128 × 128	3 × 3	32	1
MAX Pool	64 × 64	2 × 2	32	2
Conv3	64 × 64	3 × 3	64	1
MAX Pool	32 × 32	2 × 2	64	2
Conv4	32 × 32	3 × 3	128	1
FCConv1(dense)		–	64	–
FCConv2	–	–	16	–
FCConv2	–	–	2	–

5 Experiments and Simulated Results

The proposed model performance has been verified on Drishti [37], high-resolution fundus (HRF) and Refugee [38] datasets. The former "Drishti dataset has 31 normal and 70 glaucomatous (total of 101) images". All images were taken under a fixed protocol with 30° field of view, centred on the OD and of dimension 2896 × 1944 pixels each image, ground truth was collected from three glaucoma experts, referred to as Expert-1, Expert-2, and Expert-3 with experience of 3, 5, and 20 years, respectively. Next, HRF has 15 glaucomatous and 15 normal images with 3504 × 2336 image dimension, captured using a Canon CR-1 fundus camera with a field of view of 45° and different acquisition setting. At last, the Refugee dataset has 80 glaucoma and 360 non-glaucoma (total of 400) images with 1634 × 1634 image dimension. Thus, for the experimentation, we have 165 glaucoma and 406 normal images. The complete set of databases used for experimentation has been shown in Table 2.

For the assessment of our architecture performance, the data has been divided into 70% training, 10% validation, and 20% testing. Usually, when working with neural networks, evaluating results directly on the testing set can easily result into over-fitting problems, so it is better to first validate the performance on a small set for tuning the parameters and layers. Figures 6 and 7 show the loss and accuracy curve for training and validation datasets for 200 epochs while training the model. It is clearly visible that the train and validation losses decrease with epochs and

Table 2 Detailed glaucoma public datasets

S. No	Dataset name	Glaucomaa	Normal	Dimension
1	Drishti	70	31	2896 × 1944
2	Refugee	80	360	1634 × 1634
3	HRF	15	15	13504 × 2336

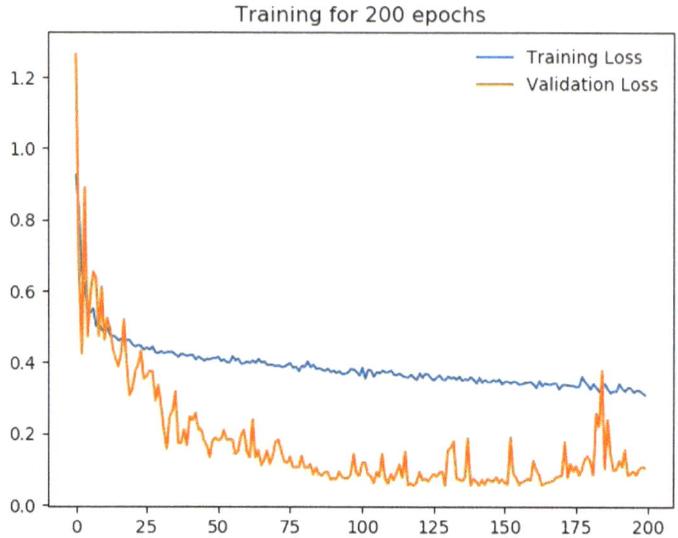

Fig. 6 Loss curve

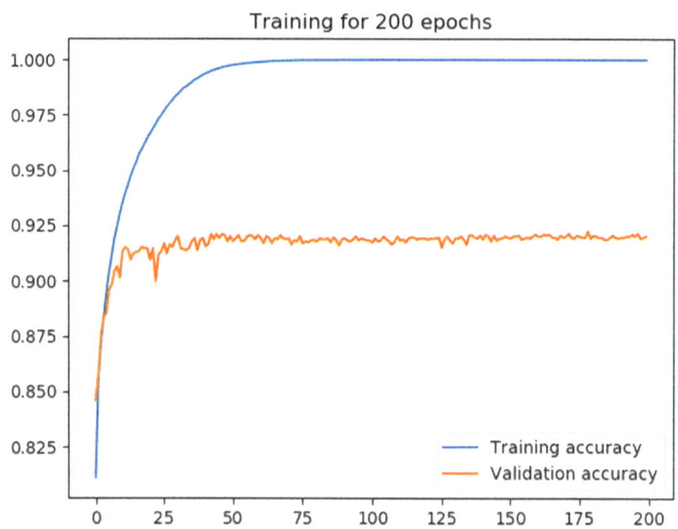

Fig. 7 Accuracy curve

converges near to each other. Although, the model suffers from small over-fitting problem, due to less variance of data, which can easily be dealt with more input data. Also, Table 3 shows the detailed classification results for each dataset individually and collectively. Table 4 shows the complete list of evaluation metrics evaluated for the proposed work.

Table 3 Detailed Results for Classification

Subject	samples used for training	samples used for testing	correctly classified samples	% age correct prediction	
Total	Normal	131	25	21	84
	Glaucoma	140	25	24	96
	Total	271	50	45	95.0
Refugee	Normal	100	10	10	100
	Glaucoma	70	10	10	100
Drishti	Normal	21	10	8	80
	Glaucoma	60	10	10	100
HRF	Normal	10	5	3	60
	Glaucoma	10	5	4	80

Table 4 Results of sensitivity, specificity, and positive predictive value classifier

TP	FP	TN	FN	Sensitivity(%)	Specificity(%)	Positive predictive value(%)	Accuracy(%)	F-1 score
24	4	21	1	96	84	85.72	90.0	91.45

5.1 Evaluation Metric

For evaluating our results, we have calculated the accuracy, sensitivity, specificity, and positive predictive rate (PPR) measures for Drishti, HRF, and Refugee datasets. Accuracy is the degree to which the calculated result matches the ground truth (standard). For defining the accuracy mathematically, we have used true positive (TP) and true negative (TN), false positive (FP), false negative (FN) values, where TP represents the images classified as glaucomatous and they are actually glaucomatous, TN is the number of images which are correctly classified as healthy, and actually, they were healthy. False positive (FP) is the healthy images misclassified as glaucomatous, and lastly, false negative (FN) is the glaucomatous samples identified as normal. The other parameters used for evaluation can be easily derived from these definitions. Sensitivity can be defined as glaucomatous category classified as glaucomatous or unhealthy. Specificity can be expressed as probability of normal being predicted as normal class. The positive predictive value (PPV) or precision defines how best one can classify each class. Also, F1-score has been evaluated which measures the test's accuracy. Figure 8 shows some of the results of predicted glaucoma score, first row is the set of glaucoma images, and second row shows the healthy images, respectively.

Thus, accuracy, specificity, sensitivity, and positive predictive value (PPV) can be defined as

$$\text{Accurcay} = \frac{\text{TP} + \text{TN}}{\text{Total no of images}}$$

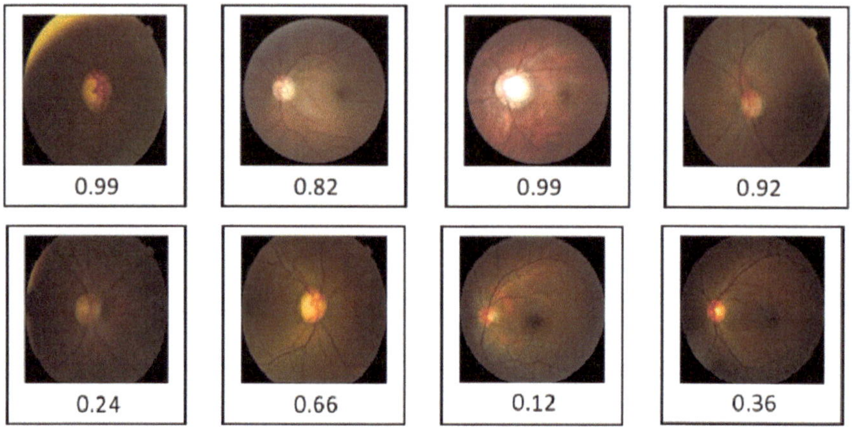

Fig. 8 Predicted glaucoma score for glaucoma and normal images

$$\text{Specificity} = \frac{\text{TN}}{\text{TN} + \text{FP}}$$

$$\text{Sensitivity} = \frac{\text{TP}}{\text{FN} + \text{FP}}$$

$$\text{Positive predictive value (PPV) or Precision} = \frac{\text{TP}}{\text{TP} + \text{FP}}$$

$$F1 - \text{score} = \frac{2 * \text{TP}}{2 * \text{TP} + \text{FP} + \text{FN}}$$

The proposed algorithm has been implemented using Keras library and NVIDIA Quadro P5000 GPU, with 128 GB RAM on windows 10 system.

6 Conclusion

Globally, the glaucoma is an irreversible eye disease that retinal disease that affects several people globally every year. Glaucoma is a chronic eye disease in which the optic nerve gets gradually damaged and thus leads to permanent loss of vision. Globally, it affects several people and thus becomes the second leading cause of eye diseases. Most of the conventional image processing techniques for glaucoma diagnosis require laborious manual inputs from a skilled eye specialist, i.e. ophthalmologists in terms of building the ground truths for the optic disc and cup boundaries. Hence, it is necessary to develop a robust and more efficient methods for glaucoma detection. Automation provides a more viable and effective solution to this problem of glaucoma detection. A proposed algorithm classifies the retinal fundus images using automated approach based on deep convolutional neural network architecture. This algorithm consists of four convolutional layers with two fully connected layers to attain accuracy of 90.0% and F-1 score of 91.45% on image classification, respectively. The proposed network has faced a over-fitting problem even after applying augmentation and dropouts. The reason is limited amount of data for the training, and thus, our future work is to collect large data to enhance the robustness of the algorithm. Apart from this, our algorithm follows a general classification problem and can be used for other medical image classification problems.

References

1. World Health Organization Media centre: visual impairment and blindness
2. Rogers K (2011) The eye: the physiology of human perception, First edn. Britannica Educational Publishing, 29 East 21st Street, New York, NY
3. Glaucoma Foundation (2012) Understanding and living with glaucoma, pp 1–36, 2012
4. Bhowmik D, Kumar KS, Deb L, Paswan S, Dutta AS (2012) Glaucoma—a eye disorder. Its causes, risk factors, prevention and medication. Pharma Innov 1(1):66–82, 2012. Retrived from www.Thepharmajournal.Com
5. Narasimhan K, Vijayarekha K (2011) An efficient automated system for glaucoma detection using fundus image. J Theor App Technol 33(1):104–110
6. Sharangouda N (2015) Automated glaucoma detection in retina from cup to disk ratio using morphology and vessel bend techniques. Int J Adv Res Comput Commun Eng 4(6):139–142
7. Anderson D (2013) The optic nerve in glaucoma. Duane's Ophthalmol., p. chap 48
8. Sivaswamy J et al (2015) A comprehensive retinal image dataset for the assessment of glaucoma from the optic nerve head analysis. JSM Biomed Imaging Data Papers 2(1)
9. Sri Abirami S et al (2013) Glaucoma images classification using fuzzy min-max neural network based on dta-core. Int J Sci Mod Eng (IJISME) 1(7). ISSN: 2319–6386
10. McIntyre R et al (2004) Toward glaucoma classification with moment methods. In: Proceedings of the first Canadian conference on computer and robot vision. IEEE
11. Almazroa A, Alodhayb S, Raahemifar K, Lakshminarayanan V (2017) An automatic image processing system for glaucoma screening. Int J Biomed Imaging vol 2017, Article ID 4826385, 19 pages
12. Almazroa A, Alodhayb S, Osman E, Ramadan E, Hummadi M, Dlaim M, Alkatee M, Raahemifar K, Lakshminarayanan V (2018) Retinal fundus images for glaucoma analysis: the RIGA dataset. In: Medical imaging 2018: imaging informatics for healthcare, research, and applications, pp 8
13. Sun W, Alodhayb S, Almazroa A, Raahemifar K, Lakshminarayanan V (2017) Optic disc segmentation for glaucoma screening system using fundus images. Clin Ophthalmol 11:2017–2029
14. Kumar BN, Chauhan RP, Dahiya N (2016) Detection of glaucoma using image processing techniques: a review. In: 2016 international conference on microelectronics, computing and communications (MicroCom). Durgapur, pp 1–6. https://doi.org/10.1109/micro-Com.2016.7522515
15. Acharya UR, Du Dua S, Du X, Sree V, Chua CK (2011) Automated diagnosis of glaucoma using texture and higher order spectra features. Inform Technol Biomed IEEE Trans 15(3):449–455
16. Society. 27th Annual International Conference of the, pp 6608–6611 IEEE, 2006
17. Krizhevsky A, Sutskever I, Hinton GE (2017) ImageNet classification with deep convolutional neural networks. Commun. ACM 60(6):84–90. https://doi.org/10.1145/3065386
18. Simonyan K, Zisserman A (2015) Very deep convolutional networks for large-scale image recognition. In: ICLR
19. Goodfellow I, Bengio Y, Courville A (2016) Deep learning. MIT Press
20. Goodfellow IJ, Pouget-Abadie J et al (2014) Generative adversarial networks. arXiv:1406.2661

21. Chen X, Xu Y, Wong DWK, Wong TY, Liu J (2015) Glaucoma detection based on deep convolutional neural network. In: 2015 37th annual international conference of the IEEE engineering in medicine and biology society (EMBC). Milan, pp 715–718. https://doi.org/10.1109/embc.2015.7318462
22. Abramoff M (2010) Retinal image analysis. Eng IEEE Rev 169–208
23. Wong DK et al (2008) Level-set based automatic cup-to-disc ratio determination using retinal fundus images in ARGALI. In: Conference in proceedings of IEEE engineering in medicine and biology society, vol 2008, no 2, pp 2266–2269
24. Liu J et al (2009) ARGALI: an automatic cup-to-disc ratio measurement system for glaucoma analysis using level-set image processing. In: Lim CT, Goh JCH (eds) 13th international conference on biomedical engineering. IFMBE proceedings, vol 23. Springer, Berlin, Heidelberg
25. Cheng J et al (2013) Superpixel classification based optic disc and optic cup segmentation for glaucoma screening. IEEE Trans Med Imaging 32(6):1019–1032. https://doi.org/10.1109/TMI.2013.2247770
26. Yin F et al (2012) Automated segmentation of optic disc and optic cup in fundus images for glaucoma diagnosis. In: Proceedings of IEEE symposium on computer-based medical *systems*
27. Juneja M, Singh S, Agarwal N et al (2019) Automated detection of Glaucoma using deep learning convolution network (G-net). I:n Multimed Tools Appl. https://doi.org/10.1007/s11042-019-7460-4
28. Xu Y et al (2011) Sliding window and regression based cup detection in digital fundus images for glaucoma diagnosis. In: Fichtinger G, Martel A, Peters T (eds) Medical image computing and computer-assisted intervention—MICCAI 2011. MICCAI 2011. Lecture Notes in Computer Science, vol 6893. Springer, Berlin, Heidelberg
29. Xu Y, Lin S, Wong DWK, Liu J, Xu D (2013) Efficient reconstruction-based optic cup localization for glaucoma screening. In: Mori K, Sakuma I, Sato Y, Barillot C, Navab N. (eds) Medical image computing and computer-assisted intervention—MICCAI 2013. MIC-CAI 2013. Lecture Notes in Computer Science, vol 8151. Springer, Berlin, Heidelberg
30. Sevastopolsky A (2017) Pattern recognit. Image Anal 27:618. https://doi.org/10.1134/S1054661817030269
31. Mvoulana A et al (2017) Fully automated method for glaucoma screening using robust optic nerve head detection and unsupervised segmentation based cup-to-disc ratio computation in retinal fundus images. Comput Med Imaging Graph 101643. https://doi.org/10.1016/j.compmedimag.2019.101643. Epub 2019 Aug 14
32. Al-Bander B, Williams BM, Al-Nuaimy W, Al-Taee MA, Pratt H, Zheng Y (2018) Dense fully convolutional segmentation of the optic disc and cup in colour fundus for glaucoma diagnosis. Symmetry 10:87
33. Diaz-Pinto A et al (2019) CNNs for automatic glaucoma assessment using fundus images: an extensive validation. J BioMed Eng OnLine. https://doi.org/10.1186/s12938-019-0649-y
34. Chen X, Xu Y, Yan S, Wong DWK, Wong TY, Liu J (2015) Automatic feature learning for glaucoma detection based on deep learning. In: Navab N, Hornegger J, Wells W, Frangi A (eds) Medical image computing an computer-assisted intervention—MICCAI 2015. MICCAI 2015. Lecture Notes in Computer Science, vol 9351. Springer, Cham
35. Abbas Q (2017) Glaucoma-deep: detection of glaucoma eye disease on retinal fundus images using deep learning. Int J Adv Comput Sci Appl (IJACSA) 8(6). http://dx.doi.org/10.14569/IJACSA.2017.080606
36. Bengio Y, Courville A, Goodfellow IJ (2016) Deep learning adaptive computation and machine learning. The MIT Press
37. Sivaswamy J, Krishnadas SR, Joshi GD, Jain M, Tabish AUS (2014) Drishti-GS: retinal image dataset for optic nerve head(ONH) segmentation. In: 2014 IEEE 11th international symposium on biomedical imaging (ISBI). Beijing, pp 53–56
38. Xu Y Chinese Academy of Sciences, China, https://refuge.grand-challenge.org/

Chapter 4
Deep Learning Based Diabetic Retinopathy Prediction of Colored Fundus Images with Parameter Tuning

Charu Bhardwaj, Shruti Jain and Meenakshi Sood

1 Introduction

Gradual progress of diabetic retinopathy (DR) leads to eye-sight loss at different severity levels. If left untreated, this prolonged situation may lead to acute blindness. Non-proliferative DR (NPDR) characterized by vitreous hemorrhage and proliferative DR (PDR) characterized by neovascularization are the two major categories of DR. Patients suffering from mild NPDR require regular screening, while appropriate laser treatment is needed for moderate/severe NPDR and PDR stage patients. Mild, moderate, and severe stages are the three further sub-categories of NPDR severity. Microaneurysms (MAs), hemorrhages (HMs), hard exudates (EXs), and cotton-wool spots (CWs) [1] retinal blood vessel abnormalities arise at NPDR stage leaking blood and fluid into the retinal surface. The automated DR grading systems involve digital retinal photography which is less laborious and cost effective comparative to the manual image grading systems [2]. To reduce the subjective interpretation and screening burdens for ophthalmologists, researchers are continuously persisting for development of flexible and reliable automated screening systems [3].

A number of traditional machine-learning-based methods are explored in the literature for DR severity grading. An enhanced CAD system is developed by Etanboly et al. for the grading of NPDR stages [4]. Segmentation approach was used to localize 12 distinct retinal layers combining shape, intensity, and spatial information. Deep

C. Bhardwaj (✉) · S. Jain
Department of Electronics and Communication Engineering, Jaypee University of Information Technology, Solan, India

S. Jain
e-mail: shruti.jain@juit.ac.in

M. Sood
Department of Electronics & Communication Engineering, NITTTR, Chandigarh, India
e-mail: meenakshi@nitttrchd.ac.in

S. Jain et al. (eds.), *Advances in Computational Intelligence Techniques*,
Algorithms for Intelligent Systems,
https://doi.org/10.1007/978-981-15-2620-6_4

fusion classification network was trained to classify normal and DR grades and assess DR grades into mild/moderate grades. A CLEAR-DR system was proposed by Kumar et al. to facilitate clinical decision support for DR [5]. Tariq et al. [6] proposed a three-stage DR detection and classification system for different DR lesions like MAs, HMs, EXs, and CWs. Filter banks were used for lesion extraction, feature for each lesion candidate are extracted which is followed by lesion classification stage.

The traditional DR diagnosis methods are replaced by neural network and nowadays, researchers have focused their work toward deep neural network-based solutions for DR diagnosis. Convolutional neural networks (CNN) have proven their worth in image interpretation and analysis in medical imaging field. A two-stage deep CNN architecture for lesion detection and DR severity grading stage was proposed by Yang et al. [7], however, Chandore and Asati [8] used a large dataset to train the CNN model for detecting DR symptoms and higher accuracy is achieved using dropout layer technique. Results obtained reveal that this approach achieves comparable performance to that of human observation by professional experts. An Inception-v3 based CNN model was proposed by Gulshan et al. [9] for DR detection and diagnosis which provides better performance. Nayak et al. [10] use 17 textural parameters for differentiating normal fundus images from non-proliferative retinopathy and proliferative retinopathy. The results obtained are validated by comparing it with grading from ophthalmic experts. Melinscak et al. [11] deployed a ten-layered convolutional network for DR classification task and maximum accuracy was achieved using small image patches. Neural networks have been used in three-class DR classification.

DR classification is of high-diagnostic relevance, hence requires better interpretation approach. Although CNN-based approaches have made significant contribution in DR diagnosis, parameter tuning and class imbalance are some gaps which still prevails in their practical implementation. In this paper, four different deep learning diabetic retinopathy (DLDR) models for DR classification using fundus images are proposed. These four models are further used to propose a novel cascaded DR detection (CDR) model. CDR model architecture utilized the combined effects of variation in filter size and number of filters employing parameter tuning.

The proposed approach overcomes the class imbalance problem by fine tuning the network parameters. The proposed approach when compared with the other pretrained CNN models and state-of-the-art methods provides significant percentage improvement validating the capabilities of proposed method.

The rest of this paper is organized as: Sect. 2 describes the methodology of CNN architecture, proposed model is discussed in Sect. 3. The obtained results are discussed in Sect. 4 followed by conclusion and the future perspective in Sect. 5.

2 Methodology

This work aims at proposing a CNN-based DR detection system for fundus image categorization into normal and DR affected retinas without any user interaction. Benchmark Method to Evaluate Segmentation and Indexing Techniques in the field of Retinal Ophthalmology (MESSIDOR) dataset consisting of 1200 fundus images is used to carry out the experiments [12]. Out of these, 800 images were acquired after pupil dilation and remaining 400 were captured without pupil dilation having standard size of 1440×960, 2240×1488 or 2304×1536 pixels at 8-bit color plane. The flowchart of proposed deep learning (DL)-based DR classification model is shown in Fig. 1. The proposed model consists of various stages involving image acquisition, data pre-processing, and augmentation and DL model for DR classification.

The proposed DL model was initially pre-trained on 1200 images until a significant level is reached so as to achieve faster classification. The pre-processed and augmented training data is provided as the iterative input to the DL model. Layers of CNN architecture are responsible for determining the best features required for classification. The increased number of layers makes the model more efficient. The lower learning rate of 0.0001 is considered for the total epoch size to improve the overall network accuracy. 70% of data is used for training phase and 30% of the data is kept for testing. During the testing phase, the model extracts imaging features from the new image and compares the feature properties learnt during the training process to classify the new image. The proposed model classifies whether the patient is suffering from disease or not which helps to distinguish between normal and DR affected fundus images.

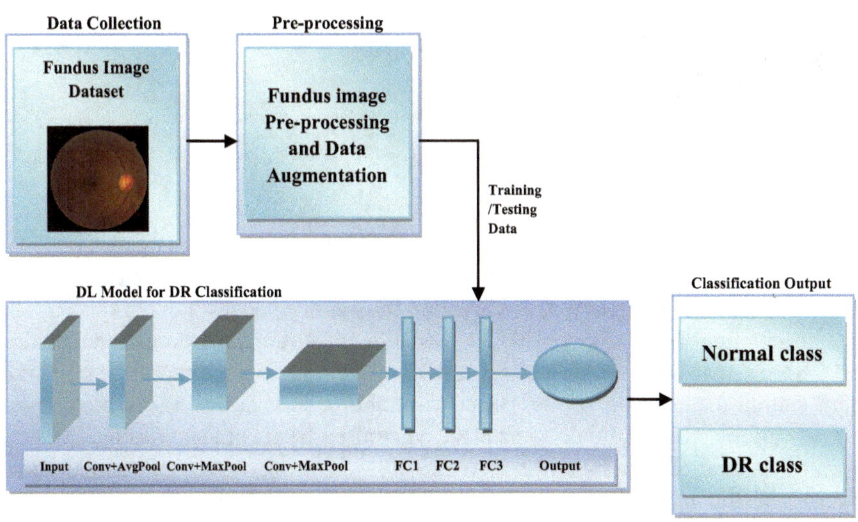

Fig. 1 Flowchart of proposed deep learning based DR classification model

Digital fundus images present in the dataset are captured from different patients at varying circumstances. Patients varying age groups, inadequate illumination, varying iris color, and several other factors may affect the inter and intra pixel intensity range creating unnecessary image variation [13, 14]. Image pre-processing steps are involved in the experimentation to remove these artefacts and further resolution normalized fundus images are resized to 128 × 128. Another important step used in the computation to improve the localization ability of network is data augmentation. Data augmentation steps utilized for this work are random rotation of 90–180°, random horizontal and vertical flips and shifts. This data augmentation step is helpful in increasing the class size as there exist limited number of training samples. The model becomes immune to variation attenuation, inadequate illumination and different orientations after applying image pre-processing and data augmentation steps [15, 16].

The major elements used in CNN architecture comprise three main layers: convolution layer, pooling layer, and fully connected layer. Convolution layer is the main element of CNN-based DL model consisting of a set of learnable filters. The dot product is computed between input and the convolution filter to produce two-dimensional maps. To introduce non-linearity, rectified linear unit (RELU) is used as it has constant gradient for positive input despite of being a non-differentiable function. Pooling layer is inserted in between the convolution layers for intermediate dimensionality reduction. Average-pooling partitions the input convolution image into non-overlapping sub-regions and outputs the average value for each of the sub-region. Max-pooling works in the same manner but returns the maximum value for each sub-region. Fully connected layer serves as the traditional neural network consisting of a large number of parameters. The complex computation in this layer, due to direct connection from every individual node to the other is resolved using dropout technique which helps in dropping out some of the nodes and the connections [17, 18].

3 Proposed Model

Authors have proposed four different DLDR models for detection of DR after reviewing the literature for other image recognition tasks. In various literatures, the model consists of stack of convolution, pooling, and fully connected layers with varying filter size and numbers. The increased numbers of convolution layers are used to make the CNN network to learn deeper features. Layers at the beginning of the network learn simple features like edges, boundaries, and curves, however, more complex features are learnt by more deeper layers. The initial layers are responsible for recognizing the edge properties in the fundus images; however, the last convolution layer learns the features for classification of fundus images into various DR grades. The learning ability of the network is improved by passing the output of convolution layer through a non-linear function and increasing its non-linear property. The layer-wise description of proposed deep learning models utilized in this paper is detailed in Fig. 2.

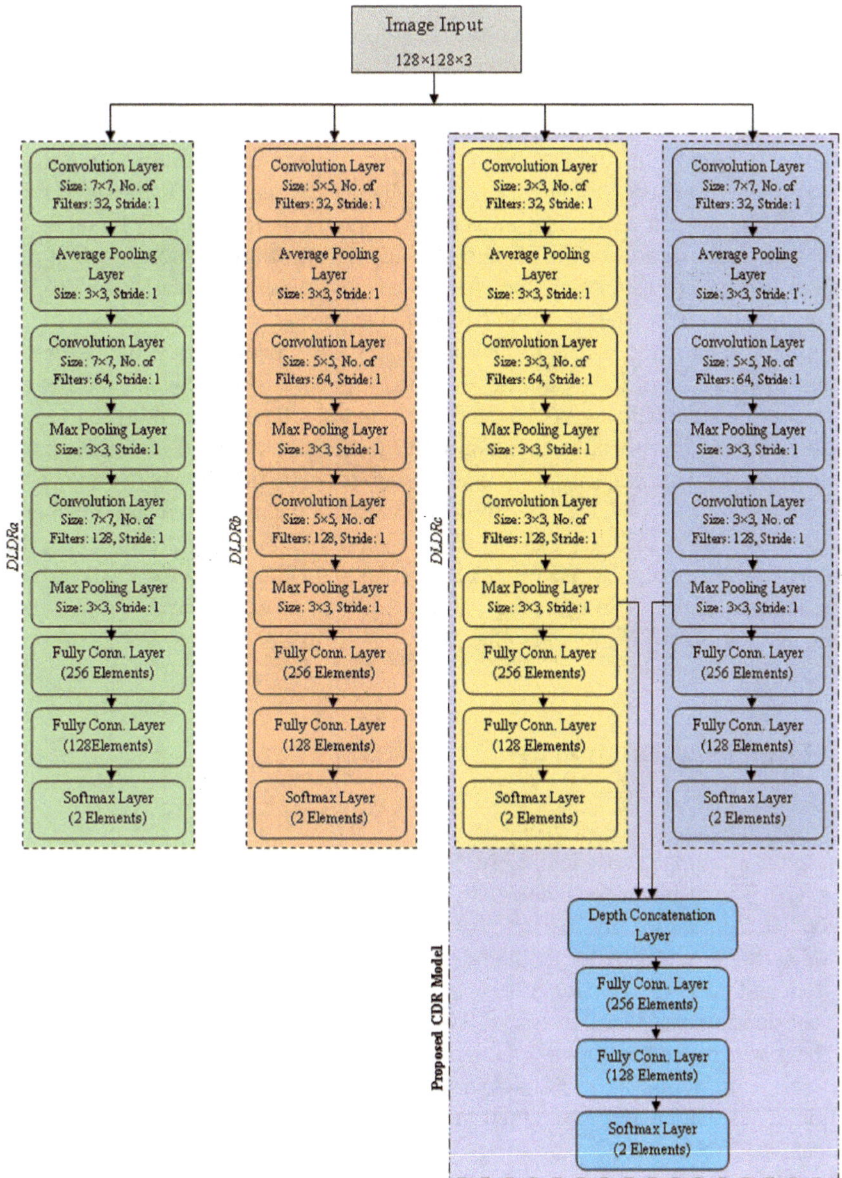

Fig. 2 Proposed architecture for CDR model

There exist various CNN models which are designated to perform different designated tasks. In this article, authors have proposed different CNN architectures to address the DR detection problem. The proposed DLDR models consist of 12 layers with alteration in convolution filter size. In DLDR model a (*DLDRa*), all the convolution layers are of size 7 × 7, *DLDRb* consists of convolution layers of filter size 5 × 5, 3 × 3 filter size is considered in *DLDRc*. A variation in the convolution filter size is the novelty of *DLDRd* with size 7 × 7 in the initial layer followed by 5 × 5 convolution filter in the next layer, and 3 × 3 filter in the last convolution layer. Pooling layers utilized in the proposed models are average-pooling and max-pooling. Both the pooling options considered in this paper uses kernel size of 3 × 3 and stride of 1 × 1. One-dimensional flattening of the CNN is accomplished using three fully connected layers that reduce the immediate dropout rate to the final classification layer. The classification output is predicted using softmax activation function and dropout is performed to reduce the chances of overfitting. By utilizing the properties of four DLDR models, a CDR CNN architecture is proposed to obtain better performance and reduced losses. The proposed CDR model is benefitted from the combined effect of variation in filter size and number of filters which are further depth concatenated before passing the features to the fully connected layers.

Conventional classification methods suffer from the problem of class imbalance by tending toward the majority class resulting into poor classification performance for the minority class. CNN has high learning capacity but the impact of class imbalance still prevails. Thus, in this paper, to address the problem of class imbalance, the same number of images is used for both normal and abnormal class. Further, data augmentation also aids in reducing the consequence of class imbalance and thus results in better classification outcomes.

4 Results and Discussion

The experimentation was performed using MATLAB2019a environment on computer system equipped with Intel Core i5 processor, 8 GB RAM, and 3 GHz processing speed. The results were analyzed by increasing the number of iterations from 1 to 200 and ranging the epoch size from 1 to 20. Keeping the epoch size and number of iterations constant for the CDR model were analyzed for its accuracy, time elapsed, and cross-entropy loss during processing with variations in terms of filter size considering base learning rate of 0.0001.

The performance parameters evaluated in this paper are accuracy, time elapsed, and cross-entropy loss [19, 20]. Accuracy provides the percentage of correct prediction of test data using the trained model and time elapsed in the time required to train the CNN model. Cross-entropy loss evaluates the performance of classification model and its value ranges in between 0 and 1 with zero (0) cross-entropy loss for perfect classification case. The results obtained for all the proposed models are tabulated in Table 1.

Table 1 Results Obtained for DLDR models

	Epoch	1	2	5	10	15	20
	Iterations	1	20	50	100	150	200
DLDRa	Accuracy (%)	33.33	40.00	53.33	60.00	73.33	75.00
	Time elapsed	00:18	00:44	00:56	01:09	01:22	01:51
	Cross-entropy loss	0.715	0.703	0.709	0.677	0.667	0.675
DLDRb	Accuracy (%)	26.67	33.33	40.00	46.67	50.00	53.33
	Time elapsed	00:10	00:35	00:47	01:01	01:26	01:35
	Cross-entropy loss	0.689	0.707	0.705	0.698	0.693	0.703
DLDRc	Accuracy (%)	13.33	33.33	46.67	50.00	53.33	60.00
	Time elapsed	01:25	01:17	01:06	00:54	00:31	00:11
	Cross-entropy loss	0.699	0.698	0.695	0.691	0.693	0.638
DLDRd	Accuracy (%)	33.31	46.76	53.34	66.67	75.08	87.51
	Time elapsed	00:08	00:29	00:52	01:04	01:16	01:23
	Cross-entropy loss	0.708	0.697	0.700	0.691	0.679	0.637

From Table 1, it is revealed that for all the models, maximum accuracies are obtained at 20th epoch and 200 iterations and later accuracy value saturates. The accuracy value for *DLDRa* is 75% at 20th epoch with 0.675 cross-entropy loss and the time elapsed for training this architecture is 1 min 51 s. *DLDRb* provides 53.33% accuracy and cross-entropy loss of 0.689 at 20th epoch consuming 1 min 35 s time which is less comparative to *DLDRc* which provides 60% accuracy with cross-entropy loss of 0.738 at 20th epoch taking 1 min 25 s time. *DLDRd* results in maximum accuracy outcome of 87.51% with cross-entropy loss of 0.637 at 20th epoch and 200th iteration consuming 1 min and 23 s of time. Out of four models proposed in this paper, *DLDRd* provides the best comparable outcomes at base learning rate of 0.0001. It can be seen that cross-entropy loss value is also minimum for this model among the other deep learning models proposed. Upon observing the outcomes beyond 20th epoch, no variations are observed in accuracy rate as well as cross-entropy loss.

This paper proposes a novel CDR model whose results were analyzed by varying the epoch size 1–20 and numbers of iterations were varied from 1 to 200 considering the base learning rate of 0.0001. The accuracy and cross-entropy values obtained for the proposed CDR model at varying epoch size are graphically represented in Fig. 3.

From Fig. 3, it can be seen that as epoch size increases from 1 to 20, accuracy value increases from 40.00% and reaches the final value of 88.63%. Cross-entropy value for the CDR model reduces from 0.693 at epoch size 1 to 0.625 at epoch size 20. The time elapsed by proposed CDR model is 3 min 41 s. The network proposed in this research work is able to learn DR classification features from the fundus images to accurately distinguish between DR infected cases from the normal cases.

The same set of processed input fundus images is also evaluated on several mainstream models; such as GoogleNet, ResNet18, AlexNet, and VggNet16 to compare

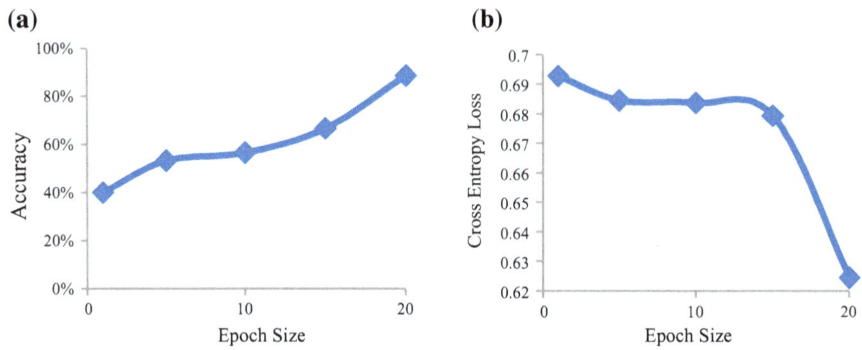

Fig. 3 Results obtained for proposed CDR model

the outcomes with the results obtained from our proposed CDR model. The comparison of the proposed CDR model with all other pre-trained models considering epoch size 20, 200 iterations, batch size 15, and constant base learning rate of 0.0001 is drawn in Table 2.

The comparison of different pre-trained model in terms of network accuracy and cross-entropy loss is depicted in Table 2. From the comparative analysis, it is revealed that the proposed CDR model yields best accuracy performance of 88.63% with cross-entropy loss of 0.6246. The comparison reveals that the proposed CDR model provides better performance with 1.13% accuracy improvement and 0.02% cross-entropy loss reduction from best available VggNet16 pre-trained model. The obtained outcome validates the proposed model parameter performance involving a trade-off with the computational cost.

Although, the DR research is mainly focused on using machine-learning approaches so far and not much development in CNN-based DR classification methods is encountered. Despite of this limitation, a comparative analysis of other DR classification approaches existing in the literature with our proposed method is presented in Fig. 4.

Figure 4 depicts that comparable results are provided by the proposed scheme without feature specific detection over a more generalized MESSIDOR dataset. Benefit of using CNN-based approach is its feasibility and robustness to be used in real-time DR classification problem.

Table 2 Comparison of proposed CDR model performance with different pre-trained models at 20th epoch

Network	Accuracy (%)	Cross-entropy loss
GOOGLENET	65.56	0.6972
RESNET18	65.83	0.6841
ALEXNET	73.33	0.6763
VGGNET16	87.50	0.6394
Proposed CDR model	88.63	0.6246

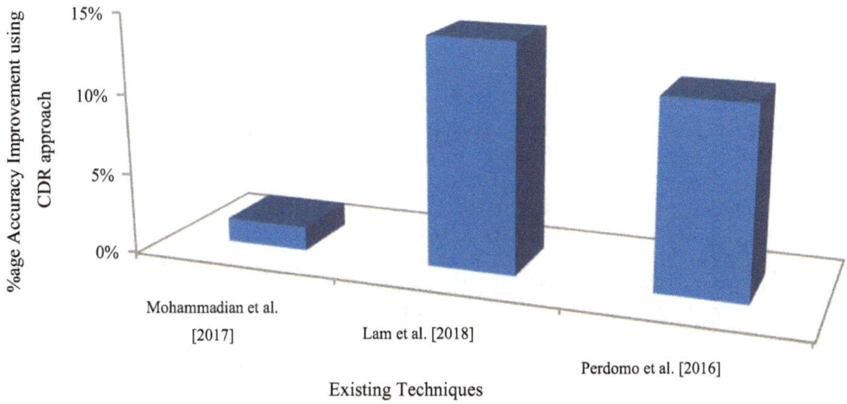

Fig. 4 Comparative analysis of CNN-based existing techniques with our proposed CDR approach

5 Conclusion

DR classification is of high clinical relevance; hence, deep learning based techniques are being explored by researchers to reduce the clinician's burden of manual retinopathy screening. Significant contribution has been made using deep learning techniques for DR diagnosis, still some gaps like parameter tuning and class imbalance problem prevail. This paper presents four different CNN-based DLDR architectures and a CDR architecture to address the problem of class imbalance by fine tuning the network parameters for DR classification. The trained CDR model provides accuracy rate of 88.63% with cross-entropy loss of 0.625 at 20th epoch providing instant diagnosis of the diseased or non-diseased fundus using single image per eye. Comparative to the best available VggNet16 pre-trained CNN model, the proposed CDR model provides 1.13% accuracy improvement and 0.02% cross-entropy loss reduction. Maximum 14.12% and minimum 0.38% improvement over state-of-the-art techniques is analyzed by the proposed approach. In the further part of this research, the CNN network will be trained to distinguish between the mild, moderate, and severe cases of DR. Moreover, the experimentation will be done on larger dataset for more subtle feature learning from fundus images. This work reveals that CNN can be trained to identify DR feature for better classification of abnormalities.

References

1. Akram MU, Khalid S, Tariq A, Khan SA, Azam F (2014) Detection and classification of retinal lesions for grading of diabetic retinopathy. Comp Bio Med 45:161–171
2. Fleming AD, Philip S, Goatman KA, Olson JA, Sharp PF (2006) Automated microaneurysm detection using local contrast normalization and local vessel detection. IEEE Trans Med Imag 25:1223–1232

3. ElTanboly A, Ghazaf M, Khalil A, Shalaby A, Mahmoud A, Switala A, El-Azab M, Schaal S, El-Baz A (2018) An integrated framework for automatic clinical assessment of diabetic retinopathy grade using spectral domain OCT images. In: 2018 IEEE 15th International symposium on biomedical ımaging (ISBI 2018), pp 1431–1435
4. ElTanboly A, Ismail M, Shalaby A, Switala A, ElBaz A, Schaal S, Gimel'farb G, El-Azab M (2017) A computer aided diagnostic system for detecting diabetic retinopathy in optical coherence tomography images. Med Phy 44:914–923
5. Kumar D, Taylor GW, Wong A (2019) Discovery radiomics with CLEAR-DR: interpretable computer aided diagnosis of diabetic retinopathy. IEEE Access 7:25891–25896
6. Tariq A, Akram MU, Shaukat A, Khan SA (2013) Automated detection and grading of diabetic maculopathy in digital retinal images. J Dig Imag 26:803–812
7. Yang Y, Li T, Li W, Wu H, Fan W, Zhang W (2017) Lesion detection and grading of diabetic retinopathy via two-stages deep convolutional neural networks. In: International conference on medical ımage computing and computer-assisted ıntervention, pp 533–540
8. Chandore V, Asati S (2017) Automatic detection of diabetic retinopathy using deep convolutional neural network. Int J Adv Res Ideas Innov Technol 3:633–641
9. Gulshan V, Peng L, Coram M, Stumpe MC, Wu D, Narayanaswamy A, Venugopalan S, Widner K, Madams T, Cuadros J, Kim R (2016) Development and validation of a deep learning algorithm for detection of diabetic retinopathy in retinal fundus photographs. Jama 316:2402–2410
10. Nayak J, Bhat PS, Acharya R, Lim CM, Kagathi M (2008) Automated identification of diabetic retinopathy stages using digital fundus images. J Med Syst 32:107–115
11. Melinščak M, Prentašić P, Lončarić S (2015) Retinal vessel segmentation using deep neural networks. In: 10th international conference on computer vision theory and applications VISAPP 2015
12. MESSIDOR (2004) Methods for evaluating segmentation and indexing technique dedicated to retinal ophthalmology. http://messidor.crihan.fr/index-en.php
13. Bhardwaj C, Jain S, Sood M (2019) Computer aided hierarchal lesion classification for diabetic retinopathy abnormalities. Int J Recent Tech Eng 8:2880–2887
14. Bhardwaj C, Jain S, Sood M (2018) Appraisal of pre-processing techniques for automated detection of diabetic retinopathy. In: 2018 fifth ınternational conference on parallel, distributed and grid computing (PDGC), pp 734–739
15. Bajwa MN, Malik MI, Siddiqui SA, Dengel A, Shafait F, Neumeier W, Ahmed S (2019) Two-stage framework for optic disc localization and glaucoma classification in retinal fundus images using deep learning. BMC Med Inform Decision Making 19:136
16. Bhardwaj C, Jain S, Sood M (2018) Automated optical disc segmentation and blood vessel extraction for fundus images using ophthalmic image processing. In: International conference on advanced ınformatics for computing research, pp 182–194
17. Rakhlin A (2018) Diabetic retinopathy detection through integration of deep learning classification framework. bioRxiv. 225508
18. Xu K, Feng D, Mi H (2017) Deep convolutional neural network-based early automated detection of diabetic retinopathy using fundus image. Molecule 22:2054
19. Mohammadian S, Karsaz A, Roshan YM (2017) Comparative study of fine-tuning of pre-trained convolutional neural networks for diabetic retinopathy screening. In 2017 24th National and 2nd ınternational ıranian conference on biomedical engineering (ICBME), pp 1–6
20. Carson Lam DY, Guo M, Lindsey T (2018) Automated detection of diabetic retinopathy using deep learning. In: AMIA summits on translational science proceedings. 147
21. Perdomo O, Otalora S, Rodríguez F, Arevalo J, González FA (2016) A novel machine learning model based on exudate localization to detect diabetic macular edema. Proceedings of the ophthalmic medical ımage analysis ınternational workshop, pp 137–144

Chapter 5
Emergency Assistive System for Tetraplegia Patient Using Eye Waver Computer Vision Technique

Tanuja Patgar and Ripal Patel

1 Introduction

The healthcare industry consists of diagnostic, preventive, remedial and therapeutic services, medical instruments, pharmaceutical manufacture and health insurance too [1, 2]. The economy chart of India shows how healthcare sector takes more participation in generation of revenue and employment. It not only is providing less universal healthcare solution but also lags far behind in world with respect to health indicators. A large part of economy of its gross domestic product (GDP) depends more on healthcare solution, but in fact it is not finding improvement.

The growth of the healthcare industry attracts lots of researchers toward it. According to statistics, the healthcare industry was the largest industry in 2006, and from 2006 to 2016, it has the highest jobs and wages and still it is expanding. The medical solutions and services have gained more attention with the increase in the population [3]. Prime reasons of growth of such industries are aging population, increase in insurance penetration, to ease in medical tourism, easy access to health and wealth centers and more importantly a key contribution of technology into healthcare industry. Increase of chronicle disease, cost of medical services, emergency room care and higher premiums are also additional factors of the growth in the healthcare sector. But without proper technology solutions, the healthcare sector cannot achieve its growth and accessibility. With better technology and innovation, the healthcare solutions become more feasible. All this can be achieved with the help of better technology and innovation.

T. Patgar (✉) · R. Patel
Electronics & Communication Engineering Department, Dr. Ambedkar Institute of Technology, Bangalore 560056, India

S. Jain et al. (eds.), *Advances in Computational Intelligence Techniques*, Algorithms for Intelligent Systems,
https://doi.org/10.1007/978-981-15-2620-6_5

1.1 Opportunities in Healthcare Industry

Health care is experiencing a new wave of opportunity.

- Big data and analytics: In the health-related industry, the number of patients is huge and medical data related to them is also huge. This type of huge data requires massive computing power and overhead costs.
- Predicting patient needs: Monitoring patient closely and predicting their risk factors in advance or needs in advance might be so much useful to the patient. Due to machine learning and big data, it is possible to predict them [4].
- Assistive technology: Many diagnostics can be done automatically using a computer vision technique to reduce the burden of doctor and aid them in the diagnostic process.

1.2 Assistive Technology

Assistive technology is one of the major factors for the disabled persons to adopt technology solution in their day-to-day activities [5, 6]. It includes biomedical products, and related accessibility can improve the lifestyle of person with disabilities, further the improvement in development and health, as well as for communication in various tracks of social life [7, 8].

Assistive technology can enhance the quality of life of disabled with all necessary comfort zones. It becomes a powerful tool to mobilize the dependence and improve the confidence of the patient. The assistive smart solution helps disabled person become independent and more social and effectively communicate in all ways of society life [9]. The participation in social activities makes them more empowers, self-esteem, greater mental function and physical balance [10].

The agenda of availability, accessibility, affordability, adaptability and acceptability factors making good strategies needs to be considered for assistive system. Nowadays, medical science is improving day by day and has developed high-tech gadgets implemented in patient's body to restore normal activities. Especially paralysis patients, such as tetraplegic patients, suffer a lot for their physical disabilities.

1.3 Tetraplegia Disease

Tetraplegia or quadriplegia is a paralysis condition where motor neuron will get damaged naturally or accidentally [11]. The patient will suffer mostly from partial or major loss of limb movement and also speech. It is difficult for these kinds of patients to communicate with their caretakers for basic needs. Many assistive smart solutions are available for bridging between patient and caretaker, but they are wearables and make the patient feel uncomfortable [12].

The biggest problem of paralyzed patients facing is independent life. When patient is in hospitals or in bed rest, it is difficult for them to communicate with caretaker and to convey their basic needs. Many existing systems are user-friendly, but the accessibility of the product becomes complex.

To fulfill both requirements, it is high time to develop a system which helps tetraplegic patients as well as people who are interested to use efficient and comfortable life [13]. Eye movement can be used by the paralysis patients and armless persons to perform simple tasks. The research proposed the acquisition and analysis of eye movements for the activation of home appliances for paralysis patients [14, 15].

In this context, Sect. 2 focuses on proposed RITA model for blink detection. Section 3 discusses the results for proposed system in detail. Conclusion is presented in Sect. 4.

2 Robust Integrated Tetraplegia Assistive (RITA) Model

Unlike conventional image processing methods for computing blinks which typically involve some combination of Eye localization, this method is using Thresholding to find the whites of the eyes and determining if the 'white' region of the eyes disappears for a period of time (indicating a blink).

We proposed robust integrated tetraplegia assistive (RITA) model, a computer vision technique to detect the blink from video in real time. The framework for the RITA model is as shown in Fig. 1.

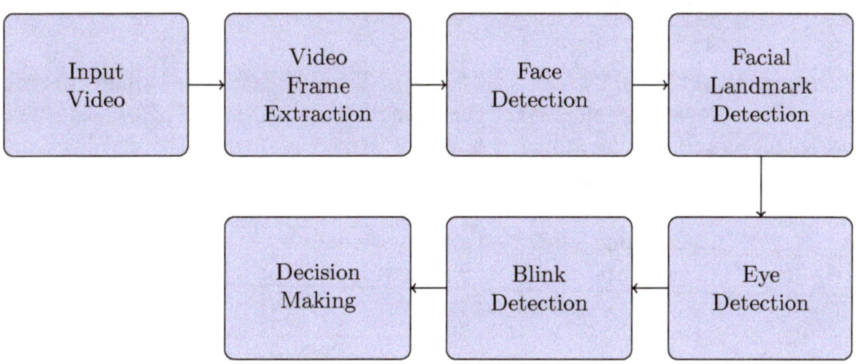

Fig. 1 RITA framework

2.1 Video Sequence Input

The system uses the camera to stream video and capture frames. It acquires a video stream from a webcam with HD resolution (1920 × 1080, 25 fps). From every frame, eye movement is detected and blinks are counted by using facial landmarks. The key facial features are highlighted using facial landmark points. Various applications are using them as bottom-up techniques to implement high-end system like face alignment, head pose estimation, face swapping and blink detection [16].

Algorithm 1 Face Detection

---

procedure FACE–DETECTION
  $I_1, I_2 \ldots I_n$ images */
  P positive images N negative images $n = P + N$
  Weight Normlization $W_{1,i} = \frac{1}{2N}, \frac{1}{2P}$
  for $t = 1, 2 \ldots T$ **do**
    Weight Updates: $W_{t,i} = \frac{W_{t,i}}{\sum_{j=1}^{n} W_{t,j}}$
    for every feature f, train classifierC_j with loss error $e_j = \sum_i W_i |C_j(x_j) - y_j|$
    Choose the classifier C_j with lowest error
    Update the weight $W_{t+1,i}$
  end for
  Final Strong Classifier: $\sum_{t=1}^{T} \alpha_t C_t(x)$
end procedure

2.2 Face Detection

The method is proposed by Viola and Jones in 2001 [17], and it is still an effective method to detect faces in images [18]. A cascade of classifiers are trained with positive and negative samples of face images. AdaBoost, a machine learning approach, is used as classifier in this method [19, 20]. In feature extraction stage, the various Haar features are calculated for positive and negative sample images. Best selection of features is achieved using AdaBoost technique. AdaBoost is weighted sum of weak classifier, which gives best classification accuracy [4]. The algorithm for AdaBoost learning has been explained in Algorithm 1.

2.3 Facial Landmark Detection

Localization and representation of important region of face are achieved using facial landmarks [21, 22]. They consist of eyes, eyebrows, nose, mouth and jawline. Face landmark detection is an exact shape prediction problem. Shape predictor tries to

Fig. 2 Facial landmark locations

adjust key point of interest along shape for a given image. Basically, it is twofold process: first, face localization, and second, key facial structure detection on the face region of interest.

Face localization is performed using Haar cascade method, and face landmark detection is achieved using Dlib's facial landmark detection. Mainly, 68 (x, y)-coordinates on face are estimated as shown in Fig. 2. For shape prediction, regression tree approach has been used. Gradient boosting optimization algorithm is utilized for learning an ensemble of regression tree [23].

The facial landmark detection is used to localize each of the important regions of face as mentioned in Eq. 1.

$$\begin{aligned}\text{FaceLandmark} &= \text{Mouth}(X_1, Y_1) + \text{RightEyebrow}(X_2, Y_2) \\ &+ \text{LeftEyebrow}(X_3, Y_3) + \text{RightEye}(X_4, Y_4) \\ &+ \text{LeftEye}(X_5, Y_5) + \text{Nose}(X_6, Y_6) + \text{Jaws}(X_7, Y_7) \end{aligned} \quad (1)$$

In the proposed work, the following landmark (as mentioned in Eq. 2) points have been used:

$$\begin{aligned}\text{FaceLandmark} &= \text{Mouth}(48, 68) + \text{RightEyebrow}(17, 22) \\ &+ \text{LeftEyebrow}(22, 27) + \text{RightEye}(36, 42) \\ &+ \text{LeftEye}(42, 48) + \text{Nose}(27, 35) + \text{Jaws}(0, 17) \end{aligned} \quad (2)$$

2.4 Eye Detection

By using facial landmarks and indexes of particular face parts, specific facial structures can be extracted [24, 25]. The main purpose here is to detect the blink; therefore, apart from eyes all parts are redundant and we focus only on eyes. Hence, from 68 (x, y)-coordinates to 6(x, y)-coordinates are used to represent an eye as shown in Fig. 3. Eye landmark position variation when eye is close and open, respectively, has been shown in Fig. 4.

For an eye detection, an easy and elegant approach is eye aspect ratio (EAR), which is based on the distance between landmark points for an eye. From Fig. 4, the relation between width and height of the coordinates is compared and EAR is calculated.

$$\text{EyeAspectRatio} = \frac{\| p_1 - p_5 \| + \| p_2 - p_4 \|}{2 \| p_0 - p_3 \|} \tag{3}$$

where p_0, p_1, p_2, p_3, p_4, p_5 are six (x, y) and landmark points used to represent the eye. The norm difference between points p_0 and p_3 represents the vertical distance which is difference between eye leads. The difference between rest of the points gives rough width of an eye which will be almost constant in any case. EAR will

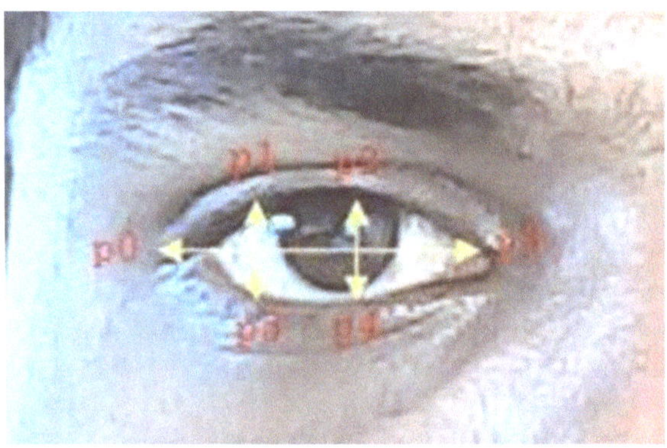

Fig. 3 Eye representation by 6(x, y)-coordinates

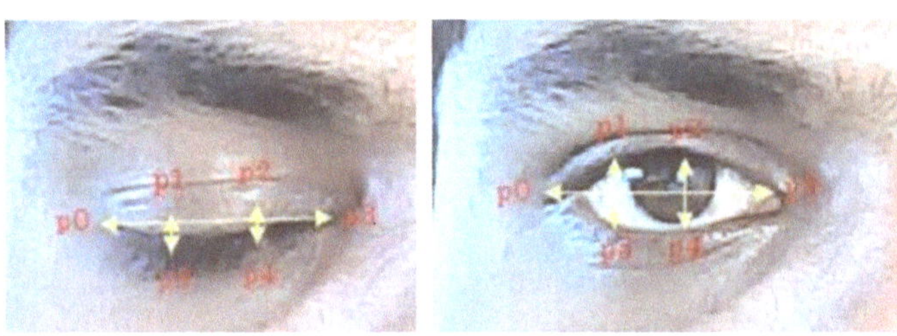

Fig. 4 Eye landmarks when eye is open (left) and close (right)

Table 1 Variation of EAR during an eye blink

Time (msec)	Aspect ratio
20	0.33
40	0.34
60	0.33
80	0.36
100	0.35
120	0.09
130	0.35
140	0.32
160	0.33

be constant when eye is open but gradually decreases to zero when it is closed. Henceforth, change in the values of EAR gives us the idea of blink detection. Table 1 shows the variation of EAR when eye is blinking.

When the eye is fully open, EAR would be larger and relatively constant over time. However, once the patient blinks, EAR decreases dramatically, approaching zero.

2.5 *Improving Blink Detector*

Sometimes due to change in illumination, face expression and lower-resolution video input, blink is detected falsely [26, 27]. To reduce such false positive detection, the temporal aspect has been considered. Over certain period of time, EAR has been calculated for each video frame and this feature vector is provided to support vector machine (SVM) classifier to detect the existence of a blink [28]. To blink detector technique more robust to these challenges, improved blink detector has been proposed as shown in Algorithm 2:

Algorithm 2 Improved Blink Detection

procedure IMP.BLINK –DETECTION
  EAR Calculation */
  for i= $N - 6$ to $N + 6$ video frames **do**
    Calculate EAR_i
  end for
  Fea_Vec=[$EAP_{N-6}EAP_{N-5}EAP_{N-4}...N...EAP_{N+4}EAP_{N+5}EAP_{N+6}$]
  input to SVM(Fea_Vec)
  Output of SVM Blink is presented or not
end procedure

2.5.1 Detecting Blinks with Facial Landmarks

After detecting the eye, the (x, y)-coordinates of the facial landmarks of the eye are recorded [29, 30]. The distance between two sets of vertical eye landmarks and the distance between horizontal eye landmarks are computed, and by combining both numerator and denominator final EAR is calculated.

Figure 5 shows the variation of EAR during an eye blink. The graph in the figure shows time versus EAR values with respect to eye movement. When the values of EAR abruptly decrease and come back to original position, it is considered as eye blink. To check the robustness of an algorithm, various face images have been used.

It is seen that EAR is almost in the range 0.32–0.36 when the eye is opened. The experimental value taken is equal to 0.33. The value rapidly drops to a value in the range 0.09–0.13 and then raises again, indicating that a blink has taken place. The

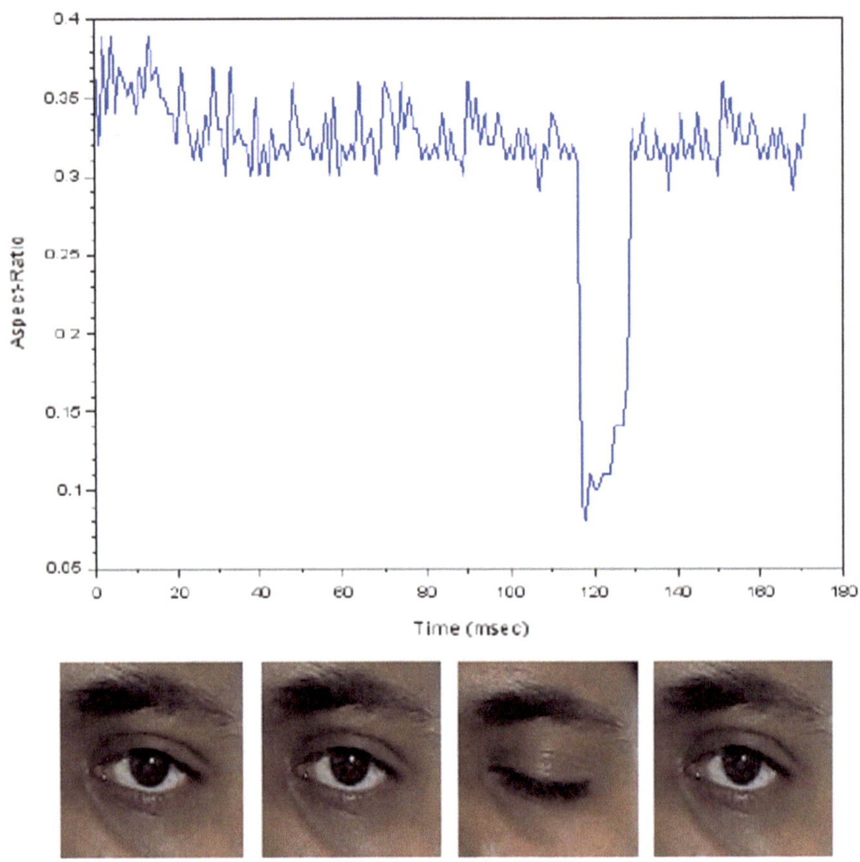

Fig. 5 Plot of EAR versus time

designed system senses the eye waver (eyeblink) using the above-mentioned values. The threshold value of the aspect ratio varies as per the patient's gaze and how wide they can open their eyes.

3 Result Analysis and Discussion

In this section, several experiments were performed to analyze the proposed predictive model. We proposed two different systems using RITA model. Both systems differ to each other in the way the combinations of outputs are generated. To assess the performance of the implemented RITA models on discriminating 2D images and to analyze various parameter effects on RITA model and a detection of blink and in result an output command, several extensive experiments are carried out.

The graph of distance between camera and eye versus detection accuracy has been presented in Fig. 6. The distance has been mentioned in inches, and detection accuracy is listed in terms of % accuracy. Till 13 in. distance, it is achieving 100% detection accuracy and detection accuracy is decreased after 13 in..

3.1 *System 1*

Based on the EAR values, eye movement and blinks, different combinations are made. These are taken as input to the system 1, and voice (sound) and switching of

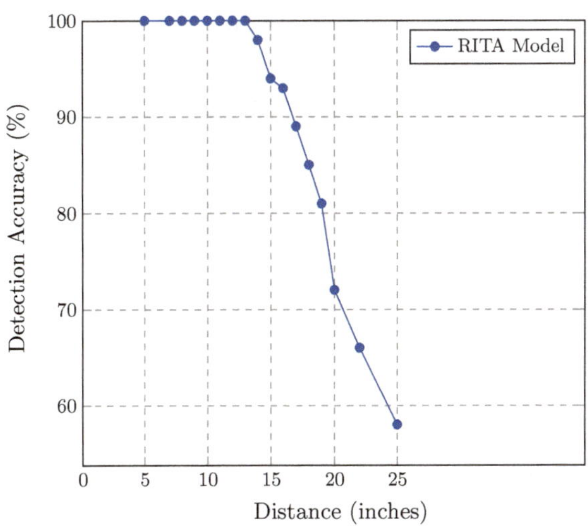

Fig. 6 Distance versus detection accuracy

Table 2 Input–output for system 1

Input combinations	Output
Left + blink	Water (voice)
Right + left + center + blink	Light ON/OFF
Right + blink	Fan ON/OFF
Eye close	Emergency

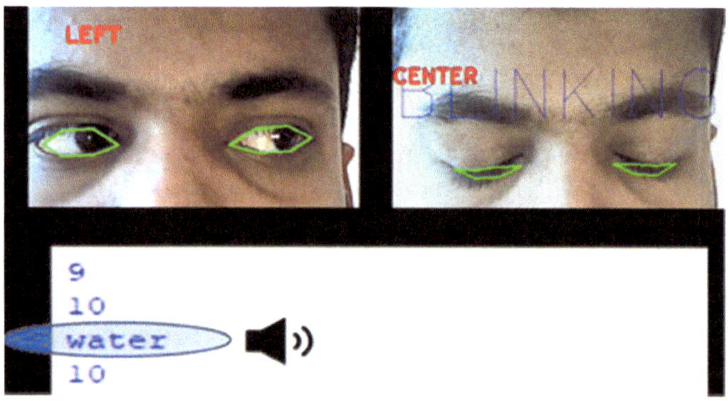

Fig. 7 Water (voice)

home appliances are achieved from the speaker as the output. The different input combinations are as shown in Table 2.

Figure 7 represents the result of system 1 when person's eyes move toward left side and then blank, which results in output of voice command in terms of 'water.' It means person is asking for water. According to this sequence, other combinations have been formed and executed and results have been shown in Figs. 8, 9 and 10. For all four combinations, system 1 is working properly and all outputs are correct.

3.2 System 2

Based on the EAR values, eye movement, head movement and blinks, different combinations are made. These are taken as input to the system 2, and voice (sound) and switching of home appliances are achieved from the speaker as the output. The different input combinations are as shown in Table 3.

Figures 11, 12, 13, 14 and 15 show the eyeball movement, head movement and blinks and their corresponding outputs from system 2. Three inputs give more combinations of the outputs for system 2. So, when patient will bend right or left, output

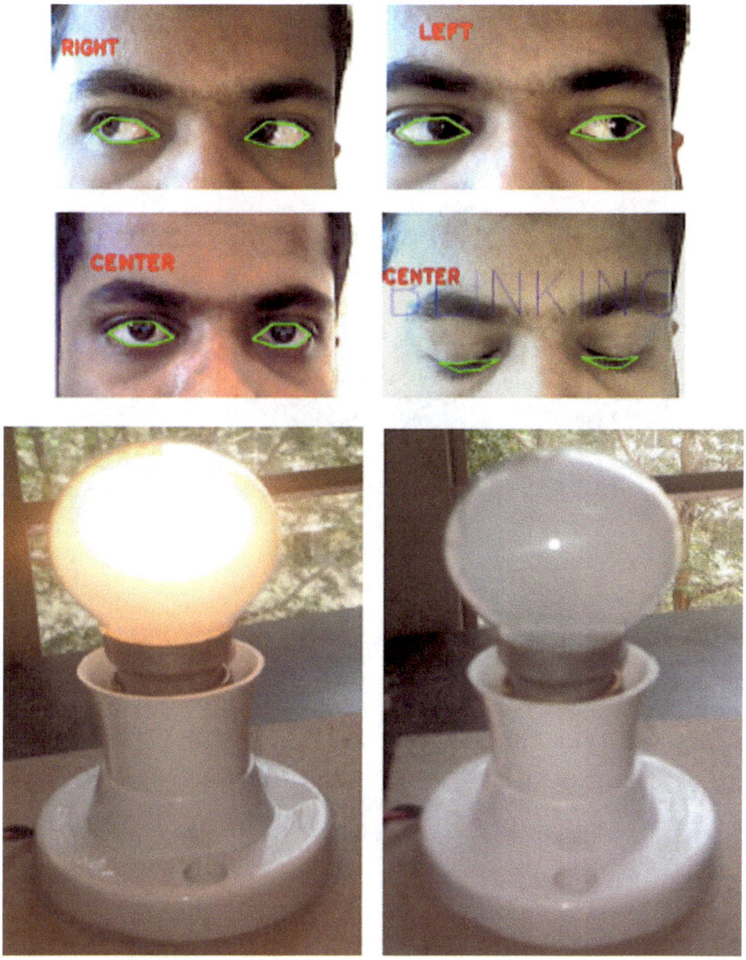

Fig. 8 Switching light ON/OFF

voice command will be generated for water and washroom, respectively. Likewise, other combinations have been generated for light ON/OFF, fan ON/OFF and emergency. Another system can be built solely based on the eye movement only. The number of eye blinking can be converted into predefined voice commands as shown in Table 4.

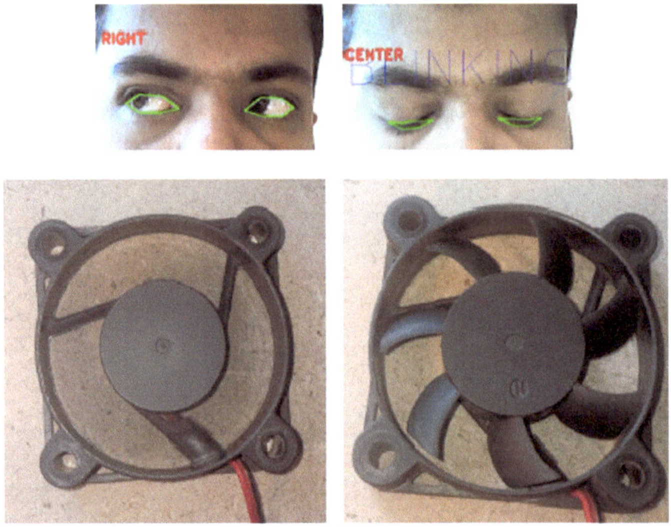

Fig. 9 Switching fan ON/OFF

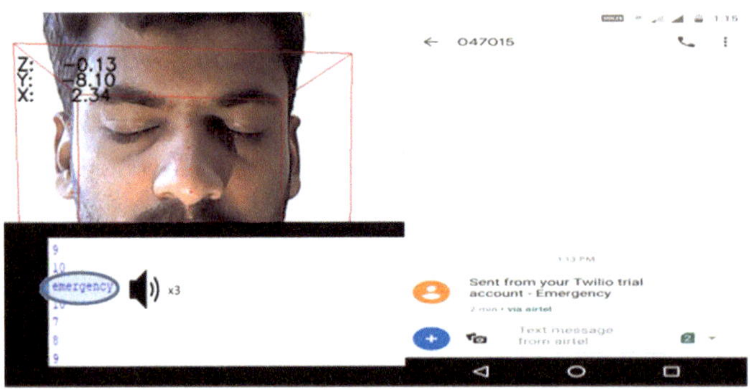

Fig. 10 Emergency

Table 3 Input–output for system 2

Input combinations	Output
Right head bend	Water (voice)
Left head bend	Washroom (voice)
Right + left + center + blink	Light ON/OFF
Right + blink	Fan ON/OFF
Eye close	Emergency

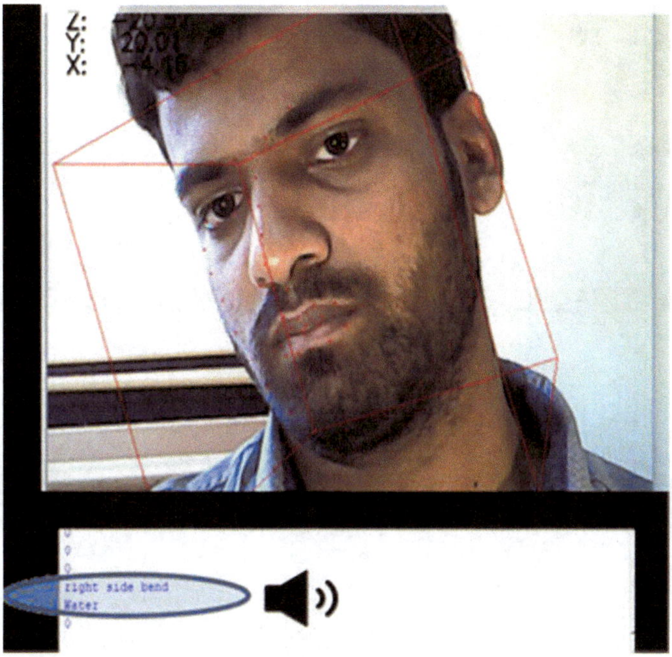

Fig. 11 Water voice

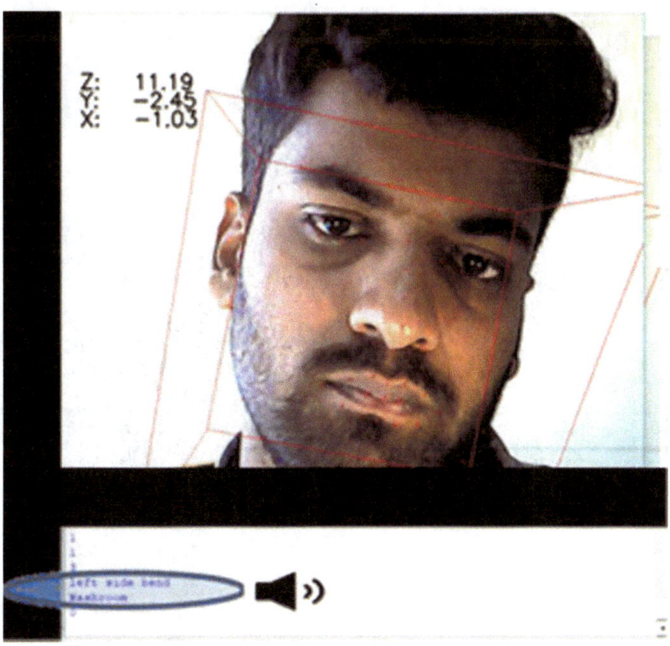

Fig. 12 Washroom (voice)

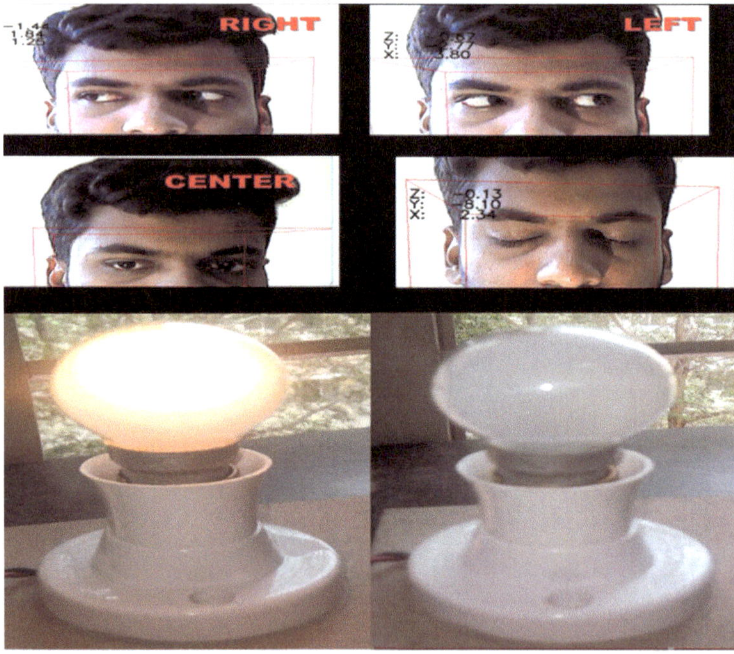

Fig. 13 Switching light ON/OFF

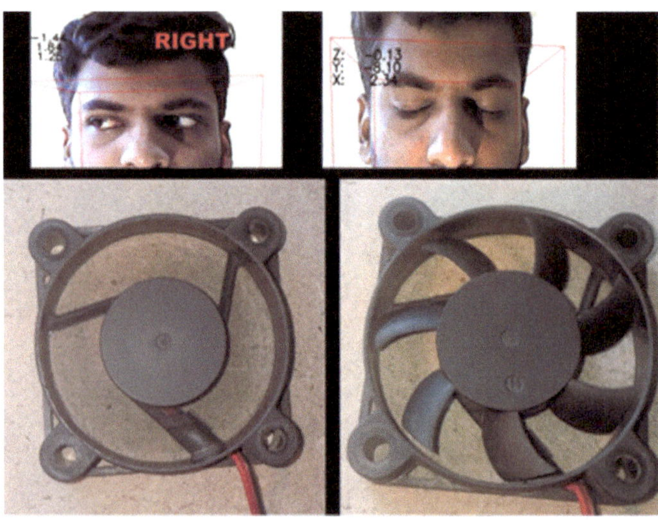

Fig. 14 Switching fan ON/OFF

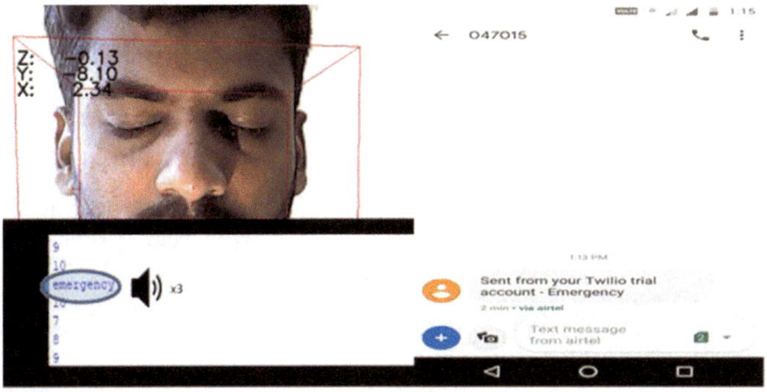

Fig. 15 Emergency

Table 4 Combination of eye wavers and its corresponding voice outputs

Number of blinks	Output voice
1	Water
2	Washroom
3	Light
4	Fan
5	Emergency

4 Conclusion

The proposed RITA systems aim to bring out a solution for the paralyzed people to communicate with the caretaker without any harm to their body externally or internally. None of the components are in direct contact with patient's body. Eyeblink detection is a very challenging problem for communication in a real-time application. This is due to the movement of the eyes and the variation of light. The proposed RITA model provides a better improvement for eye detection and blinking. The real value of eye movement and blink of the patient is recorded, processed and converted into corresponding voice output. The voice output can use the system for device automation of controlling fan, lights and basic needs. Test results show 100% detection accuracy for a distance of 13 in. Eye aspect ratio is almost in the range 0.32–0.36 when the eye is opened and value rapidly drops to a value in the range 0.09–0.13 and then raises again, indicating that blink has taken place. Artificial light is used to improve the accuracy of detection as well as the blinking for a distance equal to 13 in.

References

1. Mohammed AA , Anwer SA (2014) Efficient eye blink detection method For disabled helping domain. Int J Adv Comput Sci Appl (IJACSA) 5(5)
2. Esaki S, Ebisawa Y, Sugioka A, Konishi M (1997) Quick menu selection eye blink For eye-slaved nonverbal communicator with video-based eye-gaze detection. In: 19th international conference–IEEE/EMBS. Chicago, USA

3. Kishore Kumar G, Kemparaju N, Parineetha NS, Praveen JS, Tanuja P, Kulkarni AL (2019) Eye waver technology based assistive system for disabled. Int J Latest Technol Eng Manag Appl Sci (IJLTEMAS) VIII(V)
4. Kazemi V, Sullivan J (2014) One millisecond face alignment with an ensemble of regression trees. In: Proceedings of the IEEE conference on computer vision and pattern recognition, pp 1867–1874
5. Ayudhya CDN, Srinark T A method for real-time eye blink detection and its application
6. Lakhani K, Chaudhari A, Kothari K, Narula H (2015) Image capturing using blink detection. Int J Comput Sci Inf Technol 6(6):4965–4968
7. Muhammad A, Badruddin N, Drieberg M (2013) Automated eye blink detection and tracking using template matching. In: 2013 IEEE students conference on research and development (SCOReD), 16–17 Dec 2013. Putrajaya, Malaysia
8. Fischer JD, van den Heever DJ (2016) Portable video-oculography device for implementation in sideline concussion assessments: a Prototype. In: 2016 IEEE conference
9. Rupanagudi SR, Vikas NS, Bharadwaj VC , Manju, Dhruva N, Sowmya KS (2014) Novel methodology for blink recognition using video oculography for communicating. In: IEEE conference publications, ICAEE
10. Kiyama M, Iyatomi H, Ogawa K (2012) Development of robust video-oculography system for non-invasive automatic nerve quantification. In: 2012 IEEE-EMBS conference on biomedical engineering and sciences, IEEE conference publications
11. Jansen SMH, Kingma H, Peeters RLM, Westra RL (2010) A torsional eye movement calculation algorithm for low contrast images in video-oculography . In: 2010 annual international conference of the IEEE engineering in medicine and biology, IEEE Conference Publications
12. Karthik KP , Basavaraju S (2015) Department of electronics and communication Engineering, Sapthagiri College of Engineering, Bangalore, Design and implementation of unique video oculographic algorithm using real time video processing on FPGA. IJSRD -Int J Sci Res Devl 3(04):2321–0613. | ISSN (online)
13. Sowmya KS, Roopa BS, Prakash R Don bosco institute of technology, electronics & communication, Bangalore. A novel eye movement recognition algorithm using video oculography. In: Proceedings of international conference on recent trends in signal processing, Image Processing and VLSI, ICrtSIV
14. Naruniec J, Wiezczorek M, Kowalsak M (2017) Webcam-based system for video- oculography
15. Topal C, Gerek ON, Doğan A (2008) A head-mounted sensor-based eye tracking device: eye touch system. In: Proceedings of the 2008 Symp. On Eye tracking research and applications, pp 87–90
16. Viola P., Jones M (2001) Rapid object detection using boosted cascade of simple features. In: Proceedings of the conference on computer vision and pattern recognition (CVPR), vol 1. Hawaii, USA, pp 511–518
17. Soukupova T, Cech J (2016) Eye blink detection using facial landmarks. In: 21st computer vision winter workshop, Rimske Toplice, Slovenia
18. Adolf F (2003) How-to build a cascade boosted classifiers based on Haar-like Features. OpenCV's Rapid Object Detect
19. Hewitt R (2007) Seeing with OpenCV finding faces in images. SERVO 48–52
20. Bradski GR (1998) Computer video face tracking for use in a perceptual user interface. Int Technol J Q
21. Rachitha S, Prashanth Kumar KN (2016) Eyelid movement communication media for paralyzed patient using Fn-M16p module
22. Hewitt R (2007) Seeing with OpenCV A computer-vision library. SERVO 62–66
23. Hewitt R (2007) Seeing with OpenCV follow that face. SERVO 36–40
24. Nakanishi M, Mitsukura Y, Wang Y, Wang YT (2012) Online voluntary eye blink detection using electrooculogram. In: 2012 international symposium on nonlinear theory and its applications
25. Frigerio A, Hadlock TA, Murray EH, Heaton JT Infrared based blink detection glasses for facial pacing towars a bionic blink

26. Chau M, Betke M (2005) Real time eye tracking and blink detection with USB cameras. Boston University, USA
27. Saravanakumar S, Selvaraju N (2010) Eye tracking and blink detection for human computer interface. Int J Comput Appl 2(2):7–9
28. Lin K, Huang J, Chen J, Zhou C (2008) Real time eye detection in Vedio stream. In: Fourth international conference on natural comutational 193–197
29. Gupta A, Rathi A, Radhika Y (2012) Hands free pc control of mouse cursor using eye movement. Int J Sci Res Publ 2(4)
30. Han P, Liao JM (2009) Face detection based on adaboost. In: International conference on apperceiving computing and intelligence analysis, ICACIA 2009, pp 337–340

Chapter 6
Computer-Aided Textural Features-Based Comparison of Segmentation Methods for Melanoma Diagnosis

Khushmeen Kaur Brar, Ashima Kalra and Piyush Samant

1 Introduction

The largest organ of integumentary system that guards unrevealed muscles, bones, and internal organs in humans is skin. Skin acts as a barrier that provides protection from mechanical impacts, temperature variations, radiations, etc., and therefore its functioning has greater importance. Since the epidermis is unveiled to superficial domain, the diseases occur more to it. Abrasions are the fundamental clinical sign of various diseases like melanoma, chickenpox, etc. Skin abrasion is a division of skin that has atypical growth contrasted to skin around it. The early diagnosis of skin disease is troublesome process for most of the inexperienced dermatologists. But, it is possible to diagnose the same by incorporating digital image processing. Consequently, developing computer-aided design or CAD has become prime research area. For the time being, skin cancer has been identified as major cause of deaths. Skin cancer is standout among all cancers that accounts for about 1.6% of the total number of cancers worldwide. Its treatment requires chemotherapy and radiotherapy in the same vein as other cancer types such as breast cancer, brain tumor, and lung cancer when turned up at metastasis state. In order to refrain from traumatic procedures, early detection is unfailing solution for successful treatment. It is classified into the following categories: melanoma, basal cell carcinoma (BCC), and squamous cell carcinomas (SCC). BCC and SCC are grouped together as non-melanoma. The vast majority of skin cancers are BCC and SCC. While malignant, they are improbable to spread to other parts. BCC initiates from basal cells and SCC from squamous cells. Melanoma is one of the most dangerous cancers. About 132,000 cases of melanoma are diagnosed globally and 2–3 million accounts for non-melanoma. Melanoma arises

K. K. Brar (✉) · A. Kalra
Chandigarh Engineering College, Mohali, India

P. Samant
Chandigarh University, Mohali, India

S. Jain et al. (eds.), *Advances in Computational Intelligence Techniques*,
Algorithms for Intelligent Systems,
https://doi.org/10.1007/978-981-15-2620-6_6

from melanocytes and (malignant) is the most violent cancer that tends to spread. Malignant melanoma is augmented lesion possessing multiple colors and its borders tend to be irregular and misaligned. Although melanoma is not much prevalent; it accounts for about 5% of other skin cancers but what makes it fatal is the fact that it metastasizes unlike other skin cancers. Skin melanoma incidence rates are increasing significantly, it acquires a much attention from public health fields and research fields too. Early diagnosis of melanoma is of the utmost importance which may lead to successful treatment. The well-trained dermatologists could diagnose such cases making use of asymmetry, border irregularity, color, diameter, enlargement (ABCDE) guidelines and accomplish accuracy of 80%. Regardless of the clinical dermatologist tests being expensive, they lack in consistent standards. So there is a need of an effective and reliable automatic diagnosis method. For researchers, skin cancer detection has been a striking topic and therefore they have come up with various techniques in CAD systems to improve performance metrics such as accuracy, sensitivity, specificity, receiver operator curve, etc. Contingent upon melanoma skin injury portrayal, skin images encounter image accession, pre-refining, subdivision, feature extraction, and categorization. The objective of physicians is early spotting of melanoma to counterbalance endurance liability. For the time being, dermatologists maneuver a technique known as dermoscopy which is a non-invasive strategy for in vivo scrutiny of abrasions. The dermatologist puts gel on abrasion and scrutinizes it with augmentation instrument which augments the lesion and entitle the identification of those surfaces that are unnoticeable to naked eyes by virtue of several diagnosed algorithms like ABCD rule [1–4], Menzies method, seven-point checklist [5]. The first step for these three algorithms is the identification of lesion as melanocytic or non-melanocytic. The second step diagnosticates benign melanocytic lesion and melanoma. ABCD rule allots reckon to an abrasion and according to it, value higher than 5.45 pinpoints it as melanoma. Menzies method is specified dermoscopy method for diagnosing melanoma that is based on 11 features which are scored as present or absent. This method uses positive or negative features. For melanoma diagnosis, both the negative features should be absent. The other condition is that no less than one out of nine positive characteristics should be there. The seven-point checklist also assigns score to lesion but it only inspects the lesion for the presence of atypical differential structure. The feature score should be one and the score greater than or equal to three indicates melanoma. Advocating a smart arbitration morale booster to diagnose mild and pernicious abrasion, it put to use diverse stages including pre-refining such as clamor evacuation, segmentation and feature extraction from abrasion area, feature-excerption, and categorization. Genetic algorithm can be employed to conduct feature optimization. So numerous abrasion categorization methodologies have been enlisted like motif codification of dermoscopy portrayals for perceptually monotonous framework as explained by Abbas et al. [6]; methodologist approach for classification of pigmented skin lesions as explained by Celebi et al. [7], transformation approach segmentation algorithm for identification of abrasion region not beyond an ellipse as explained by Yuan et al. [8] and Rajab et al. [9] proposed two methods namely ‘region-based subdivision tactic’ and another 'edge detection' method. Edge detection method is neural network-based approach. 'Detention basin-based algorithm for self-activating subdivision' as explained by Wang et al. [5], 'Random walker algorithm' for automatic segmentation as explained by Wighton et al. [10], '3D ambit statistics to RGB color images to boost subdivision of tinted and non-tinted abrasion' as explained by Li et al. 'Combination of classical JSEG algorithm and particular pattern operator' as explained by Komati et al. [11], 'automated method for melanoma recognition using texture analysis' as explained by Sheha et al. [12]; 'cybernetics stratification procedure employing ANN classifier' as explained by Aswin et al., 'self-activating skin cancer stratification procedure using semantic net classifier with wavelet and curvelet facets' as explained by Mahmoud et al. [13]. Notwithstanding innumerable classifiers happen to be wielded; the amalgamation of legion classifiers is bagging significance on the grounds that furtherance in classification can be procured. Nevertheless, most people diagnosed with skin cancer have higher likelihood of healing; melanoma survival rates are lower than those of non-melanomas. As skin cancer cases are progressively increasing, an automated system for early diagnosis is immensely prominent. As proclaimed, dermoscopy increases sensitivity of melanoma by 10–27%, it only increases diagnostic performance if dermatologist is edified. Maneuvering diagnostic algorithm is least preferred as diagnosis by dermatologist depends upon human vision and experience. So automated dermoscopy image analysis can be used as diagnostic appliance for investigating skin lesion. There is a huge variety of image segmentation algorithms. Foregoing research works divided the same into three main categories: 1. Statistics and probabilistic which is for high contrast images and does not work well for overlapped regions, 2. Machine learning that makes use of statistical techniques to make computer 'Learn' data and robust segmentation, and 3. Active contour. In case of skin cancer, lesion area might really be irregular with undefined curves that need to be segmented accurately as whole of the skin cancer will not be cloistered if subordinate segmentation is performed. And in case, adequate segmentation is performed, some flesh might cut off. Consequently, algorithms in active contour model are dealt with to decipher such difficulties.

2 Literature Review

From the past decade diverse ideologies relic for the recognition of melanoma; part of which attempt to imitate the performance of medical specialists by diagnosing and concentrating various dermoscopic attributes. The attributes conceivably count an abrasion in the same way, the dermatologists endorse. Various organizations have advocated CAD networks for melanoma recognition. The network wields various characteristics, for instance, color, structure, appearance, etc., to identify images.

Currently, the neural language processing network is bagging significance to discriminate threatening abrasions from gentle ones. The recognition testament of artifact in image would apparently originate with image pre-refining such as clamor evacuation, followed by, subdivision, characteristics extraction, feature-excerption and categorization. The latest researches reveal that the amalgamation of legion classifiers is attaining importance.

Sumithra et al. [2] also ascertained self-regulating subdivision approach for abrasion categorization. The paper also elicited abrasion portion by color and appearance traits. Abrasion portrayals were pre-refined and Gaussian regularization was administered to eliminate clamor. To amplify refined image, morphological disintegration and expansion were utilized. Region expanding subdivision approach was administered. To measure the color existing in abrasion, four census from fragmented abrasion area over discrete mediums of six distinct color ranges were outperformed. For texture existing in abrasion, a set of demographic consistency descriptors contingent on GLCM were administered. To measure pursuance of fragmentation, MOL, MUS, MOS, and ER were evaluated maneuvering SVM and KNN. Pursuance was evaluated concerning accuracy, precision, recall, and F-measure. Maximum pursuance was evaluated by the amalgamation of SVM with KNN.

Bumrungkun et al. [6] ascertained image subdivision strategy contingent upon SVM and the snake active profile. The paper revealed the aspects of arrangement suitable for image and discovering margins.

Barata et al. [14] ascertained dual methodology for melanoma discernment. The first approach wielded global strategy to segregate abrasions. The subsequent approach utilized of local traits and bag-of-features classifier. This paper intended to determine the optimal methodology for abrasion categorization and to deduce the best trait by the whole of color and appearance traits through contrast. The paper practiced two gradient histograms. For color traits, six color ranges were maneuvered and the paper typified color dissemination in abrasion area employing a set of three color histograms. The BoF technique constituted composite entities by assembling local lesion designs. The categorization of local traits was accomplished maneuvering dictionary of visual words and histogram was assembled. The outcomes were performed on 176 portrayals from index of Hospital Pedro Hispano, Matosinhos. The result revealed that the color traits surpass the appearance traits. The susceptibility of 96%, selectivity of 80%, for the global methodology was achieved. For local ones, the susceptibility of 100% and selectivity of 75% were achieved when performed on 176 dermoscopic imagery.

In yet another paper, Barata et al. [15] proposed some changes to the previous work for the sake of advancement in dermoscopy images. The paper ascertained four algorithms that adopted color stats to conjecture color of luminary. The stratification composition incorporated detachment of abrasion from vigorous skin maneuvering standard segmentation. The BoF technique was trained and examined and the portrayals were partitioned employing Harris-Laplace core locator. K-means congregate was employed. Histogram of visual words was catalogued maneuvering SVM. 482 portrayals were screened from ERDA index. The paper trained five divergent BoF methodologies. three metrics were computed to scrutinize the accomplishment specifically: sensitivity, specificity, and accuracy. The calculations revealed increment in susceptibility of BoF approach to 79.7% and the selectivity to 76% maneuvering 1-D RGB histogram.

Abuzaghleh et al. [16] ascertained self-regulating abrasion subdivision and evaluation for prior diagnosis and prohibition in regard to color and appearance. The paper quantified global threshold to transfigure potency portrayal to binary one and contrasted two kinds of classifiers specifically 1-level and 2-level. The algorithm to subdivide gray image was contingent upon active contour. The foremost classifier categorized the ordinary, divergent, and melanoma portrayals with reliability of 90.3, 92.1, and 90.6%. The subsequent classifier categorized images with reliability of 90.6, 91.3, and 97.7%. The results of the subsequent classifier were better. The demonstration was dispatched on PH2 dermoscopy directory.

Jain[a] et al. [17] ascertained cybernetics methodology for the diagnosis of melanoma. The paper inputted portrayal that reckoned melanoma abrasion. Unmanned thresholding and periphery diagnosis were maneuvered for subdivision. Geometric traits were allocated and the extricated traits labeled abrasion as mortal and immortal. Subdivision was performed maneuvering Otsu thresholding. The characteristic values extricated in feature extraction stage contrasted and abrasion was classified as melanoma or typical skin or mole.

Yan et al. [18] ascertained smart arbitration morale booster to recognize mild and pernicious abrasions with the help of GA-based character inflation. The paper advocated pre-refining like dull razor and median filters for hair elimination and clamor evacuation. The portrayals were fragmented maneuvering pixel inhibition to segregate abrasion from background. GA was implemented for feature-excerption. SVM was maneuvered for mild and pernicious abrasion categorization. Edinburgh Research & Innovation index for network estimation was considered. The outcomes revealed that the framework attained propitious implementation with average of 92% and 84%.

Adjed et al. [19] ascertained amalgamation of organizational and textural characteristics from two descriptions. The paper ascertained organizational traits attained from first and second level of wavelet and curvelet coefficients maneuvering LBP operator. The optimum outcome was executed by amalgamation of wavelet and LBP^{U2}. The attained outcomes were also endorsed maneuvering random exemplification cross-authentication accompanying SVM with kernel essence on PH2 index. The framework revealed considerable outcomes concerning reliability of 95% and the susceptibility of 90% was acquired.

Monisha et al. [20] ascertained categorization of pernicious and mild abrasion employing BPN and ABCD methodology. The paper exhibits acceptable conclusions with reliability of 95% and susceptibility of 90%.

Oliveira [21] presented a computational strategy for diagnosing pigmented abrasion in comprehensive portrayals. The procedure was contingent upon an anisotropic dispersion refiner, an effective contour representation lacking fringes and a SVM. The strategy proposed by the paper attained acceptable fragmentation results, with clamorous portrayals, but not on portrayals with very low variance peripheries. Feature and abrasion categorization conferred remarkable results. The paper offered to introduce novel strategies concerning fragmentation and categorization of PSLs, for preferable detection.

Pennisi et al. [22] presented a smart and perfectly self-regulating algorithm for abrasion fragmentation. The paper employed Delaunay triangulation to extricate binary obscure. The observation was wielded on openly accessible index. The outcomes revealed that the strategy used by the very paper was immensely precise coping with mild abrasions, while the fragmentation accuracy notably diminished when melanoma portrayals were handled. The categorization evaluation attained a sensitivity of 93.5% and the specificity of 87%.

3 Methodolgy

The propounded procedure for melanoma recognition is presented in the figure. This paper advocated an image processing-based framework to perceive, extricate, and divide the abrasion from dermoscopy portrayals. The input for the framework is the image of an abrasion. The image is subsequently pre-refined to intensify the caliber (Fig. 1).

The segmentation is based on thresholding (otsu's) is employed for image subdivision. The segregated images are subsequently handed over to feature extraction and feature selection block. The extricated characteristics are subsequently transferred to classification block that systematized the abrasion into class 1 and class 2.

A. *Image acquisition*

The images were drawn from PH2 index [23]. The database holds an unabridged 200 dermoscopic portrayals of melanocytic abrasions. PH2 index is accessible for public. Most of the papers maneuvered this open PH2 database. Image acquisition is introductory step in the framework progression and in the absence of which, no additional operation is feasible.

B. *Pre-refining*

The pre-refining is a moderation of picture statistics that annihilates undesirable malformations and intensifies characteristics to allay further processing. The pre-refining incorporates subdivision techniques. Image subdivision is accomplished maneuvering thresholding by Otsu, region growing and region merging segmentation.

The foremost requisite of segregation is to alienate the intent from background. Earlier studies have proffered distinct classifications of subdivision procedure suchlike clustering-based, edge-based, etc., in this research, the subdivision based on

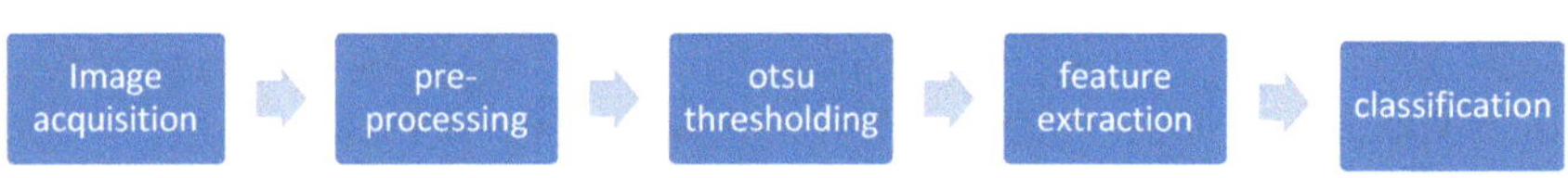

Fig. 1 Proposed technology

thresholding (Otsu's) is espoused. Abrasion segmentation is an indispensable step before feature-extrication to efficaciously segregate the abrasion as malignant or benign.

a. *Otsu Thresholding*

Otsu thresholding [24] is an expedient of automatically detecting an ideal threshold hinging on the detected organization of the pixel values. In this technique, one pursues for a threshold that curtails the intra-class dissimilarity.

C. *Feature Extraction and Feature Classification*

In order to demarcate between each input composition, feature-extrication is employed. This is the pre-eminent step which substantially regulates the pursuance of the classifier. Features extricated from GLCM [25] comprises: auto-correlation, cluster-prominence, energy, entropy, homogeneity, information measure of correlation 1, information measure of correlation 2, inverse difference, maximum probability, sum average, sum entropy, sum of square variance, and sum variance. In order to ascertain the optimal subdivision approach, and to ameliorate the efficiency, the feature selection is a central errand.

4 Result and Discussion

Classification algorithms predict a decision from the labeled feature set. These algorithms infer the hypothesis which helps in predicting the labels from the testing data set by analyzing the training dataset. Effectiveness of classifiers depends on the nature of dataset, complexity and application area. In the present study, different classifiers have been investigated, named as Complex Tree, Medium Tree, Simple Tree, Linear SVM, Quadratic SVM, Cubic SVM, Fine Gaussian SVM, Medium Gaussian SVM, Coarse Gaussian SVM, Fine KNN, Medium KNN, Coarse KNN, Cosine KNN, Cubic KNN, Weighted KNN, Boosted Trees, Bagged Trees, Subspace Discriminate, Subspace KNN and RUS Boosted Trees. These classifiers belong to different fields of statistics and computer science, and it is very difficult to understand and underlying each and every parameter of classifiers, and manual parameter selection can lead biasing in the performance index. Hence, to get the optimal performance of classifiers, parameter selection was done through cross-validation. Testing data have been trained through repeated iteration technique and validation is performed through ten-fold cross-validation. We administered six performance metrics namely sensitivity (SE), specificity (SP), accuracy (ACC), Jaccard index (JA), dice co-efficient (DI), and the divergence value (DV). Bio-medical imaging is regularly used as handy source of information for clinical study. It is a non-contact, non-invasive, and cost-effective diagnostic tool . Figure 2 shows the performance of six metrics.

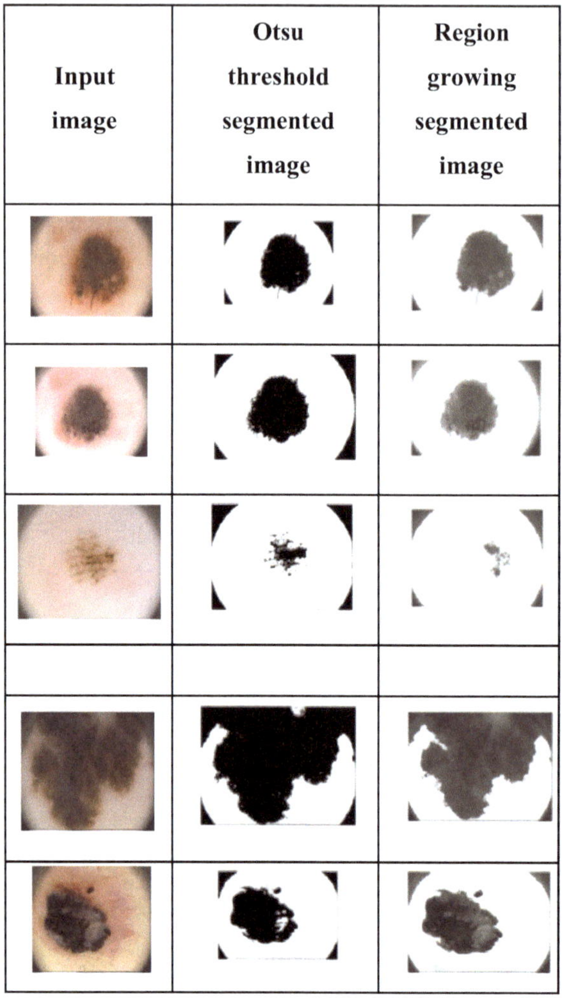

Fig. 2 Input images versus Otsu threshold segmented and region growing segmented images

A. *Jaccard Index (JA)*: The Jaccard index, otherwise known as intersection over union, and Jaccard similarity is a data that is employed for determining the resemblance and the variation of sample sets. The Jaccard co-efficient computes resemblance among finite sample sets. It is characterized by the size of the intersection upon size of the union of samples.

The Jaccard index can be stated as:

$$\mathbf{JA = TP/TP + FN + FP}$$

where

TP true positive
TN true negative
FP false positive
FN false negative

A. *Dice Co-efficient* (*DI*): The dice co-efficient is a data maneuvered to determine the resemblance of two samples. The dice co-efficient can be stated as:

$$\mathbf{DI = 2 * TP/2 * TP + FN + FP}$$

Divergence Value (*DV*): The divergence value computes the percentage of error or the percentage of segmentation errors. The divergence value can be stated as:

$$\mathbf{DV = FP + FN/TP + FN}$$

B. *Accuracy*: Accuracy is defined as the division of calculations our model got precise. The accuracy can be stated as:

$$\mathbf{Accuracy = TP + TN/TP + TN + FN + FP}$$

C. *Sensitivity* (*SE*): Sensitivity is used to gauge the performance of a categorization model. The sensitivity is calculated as:

$$\mathbf{SE = TP/TP + FN}$$

D. *Specificity* (*SP*): Specificity can be calculated by:

$$\mathbf{SP = TN/FP + TN}$$

E. Area Under the curve (AUC) analysis
AUC has the ability to access the performance of the classifier over its entire operating range. Figure 3 depicts the ROC curve between false positive rate and true positive rate of various classifiers against the selected feature set (Figs. 4 and 5).

5 Conclusion

Automatic diagnosis of skin cancer is feasible and achievable through the usage of well-defined segmentation and classification technique. While many success has been recorded in the current advances in automation of medical diagnosis, this study tends to maximize the large availability of ubiquitous devices and elicitation of past

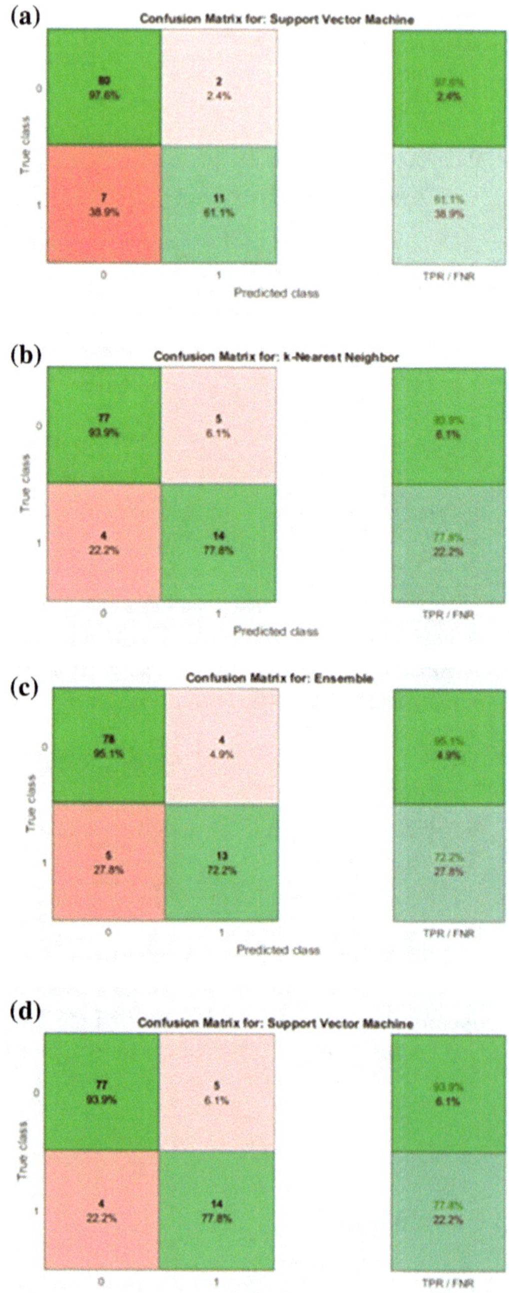

Fig. 3 Confusion matrix **a** Quadratic SVM, **b** Fine Gaussian SVM, **c** Cosine KNN, **d** Medium Gaussian SVM

Classifier	Dice Coefficient	Jaccard Index	Divergence Value	Accuracy	Sensitivity	Specificity
Complex Tree	91.01%	83.51%	17.64%	85%	89.41%	60%
Medium Tree	91.01%	83.51%	17.64%	85.00%	89.41%	60.00%
Simple Tree	89.28%	80.65%	20.93%	82.00%	87.20%	50.00%
Linear SVM	88.48%	79.34%	23%	81	87.95%	47.05%
Quadratic SVM	94.94%	89.53%	11.11%	91.00%	95.06%	73.68%
Cubic SVM	92.25%	86.20%	15.00%	88.00%	93.75%	65.00%
Fine Gaussian SVM	94.67%	89.88%	10.30%	91.00%	91.95%	84.61%
Medium Gaussian SVM	92.22%	85.55%	15.29%	87.00%	90.58%	66.66%
Coarse Gaussian SVM	90.10%	82%	18.00%	82.00%	82.00%	0/0
Fine KNN	92.02%	85.22%	16.04%	87.00%	92.59%	63.50%
Medium KNN	93.90%	88.50%	12.19%	90.00%	93.90%	72.22%
Coarse KNN	90.10%	82%	18.00%	82.00%	82.00%	0/0
Cosine KNN	94.47%	89.53%	11.11%	91.00%	95.06%	73.68%
Cubic KNN	93.82%	88.37%	12.50%	90.00%	95.00%	70.00%
Weighted KNN	93.82%	88.37%	12.50%	90.00%	95.00%	70.00%
Boosted Trees	94.54%	89.65%	10.84%	91.00%	93.97%	76.47%
Bagged Trees	92.59%	86.20%	15.00%	88.00%	93.75%	65.00%
Subspace Discriminant	90.79%	83.14%	18.51%	85.00%	91.35%	57.89%
Subspace KNN	89.15%	80.43%	21.42%	82.00%	88.09%	50.00%
RUSBoosted Trees	91.61%	86.74%	17.80%	87.00%	97.26%	59.25%

Fig. 4 Classifier performance

skin cancer diagnosis image set toward providing cost-effective, easier, and faster diagnosis for underserved areas. Six performance metrics, i.e., sensitivity, specificity, accuracy, Jaccard index, dice co-efficient, and divergence value practically accomplished good segmentation results. The imposition of these metrics on PH2 dataset images acquired DV of 10.3% for Fine Gaussian SVM, DI of 94.67% for Fine Gaussian SVM classifier, and JA with 89.88% for Fine Gaussian SVM classifier as well. The imposition provides the promising result with sensitivity of 97.26% for RUS Boosted Tree. The specificity result with 84.61% for the Fine Gaussian SVM Classifier. The accuracy of 91% is achieved for four classifiers that are Fine Gaussian SVM, Boosted Trees, Quadratic SVM, and Cosine KNN classifiers. From the results, it may be concluded that the Fine Gaussian SVM classifier outperformed the other classifiers among all the performance metrics.

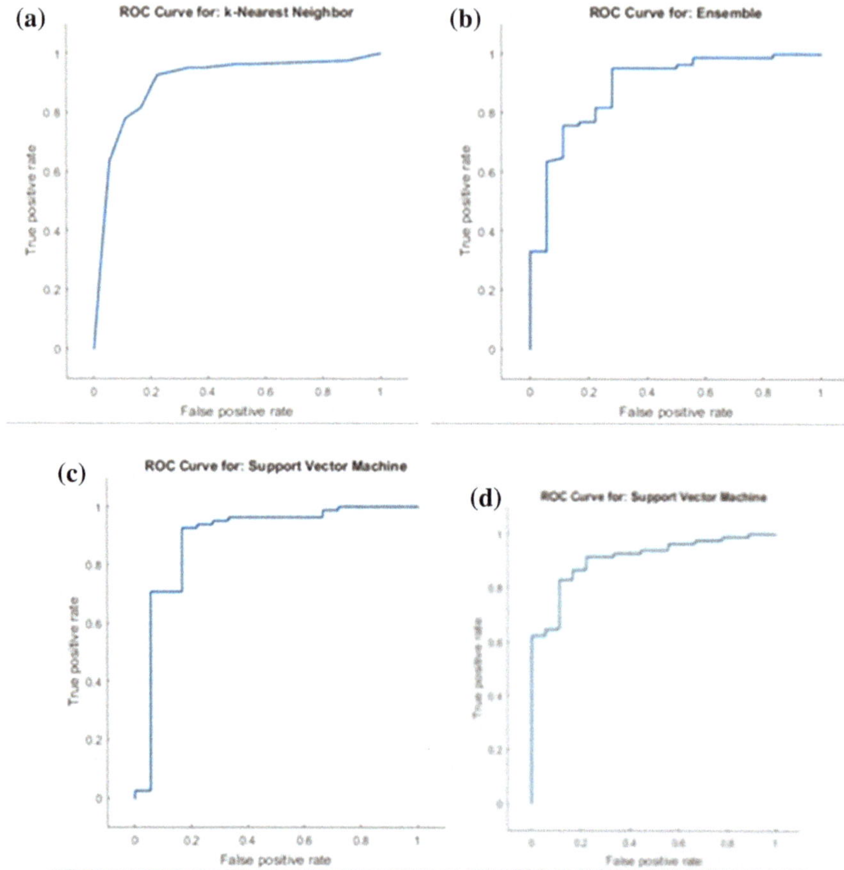

Fig. 5 Area under the curve for few classifiers **a** Cosine KNN, **b** Boosted Trees, **c** Quadratic SVM, **d** Medium Gaussian

References

1. Azmi NFM, Sarkan HM, Yahya Y, Chuprat S (2016) ABCD rules segmentation on malignant tumor and Benign skin lesion images. In: Proceedings of 2016 3rd international conference on computer and information sciences ICCOINS 2016, pp 66–70
2. Sumithra R, Suhil M, Guru DS (2015) Segmentation and classification of skin lesions for disease diagnosis. Procedia Comput 45(C):76–85
3. Okuboyejo DA, Olugbara OO, Odunaike SA (2013) Automating skin disease diagnosis using image classification. II:23–25
4. U. Dorj, K. Lee, J. Choi, and M. Lee, "The skin cancer classification using deep convolutional neural network," pp. 9909–9924, 2018
5. Kharazmi P, AlJasser MI, Lui H, Wang ZJ, Lee TK (2016) Automated detection and segmentation of vascular structures of skin lesions seen in Dermoscopy, with an application to basal cell carcinoma classification. IEEE J Biomed Health Info 21(6):1675–1684

6. Bumrungkun P, Chamnongthai K, Patchoo W (2018) Detection skin cancer using SVM and snake model. In: 2018 international workshop on advanced image technology (IWAIT), pp. 1–4, 2018
7. Emre Celebi M, Wen Q, Hwang S, Iyatomi H, Schaefer G (2013) Lesion border detection in dermoscopy images using ensembles of thresholding methods. Ski Res Technol 19(1):1–7
8. Permissions F (2017) Impulse-noise resistant color-texture classification approach using hybrid color local binary patterns and Kullback—Leibler Divergence," no. iii
9. Rajab MI, Woolfson MS, Morgan SP (2004) Application of region-based segmentation and neural network edge detection to skin lesions. Comput Med Imaging Graph 28(1-2): 61–68
10. Wighton P, Sadeghi M, Lee TK, Atkins MS (2009) A fully automatic random walker segmentation for skin lesions in a supervised setting, pp 1108–1115
11. Komati KS, Salles EO, Sarcinelli Filho M Fractal-JSEG : JSEG using an homogeneity measurement based on local fractal descriptor
12. Sheha MA, Mabrouk MS, Sharawy A (2012) Automatic detection of melanoma skin cancer using texture analysis. Int J Comput Appl 4220(20):975–8887
13. Mahmoud MKA, Al-Jumaily A (2011) The automatic identification of Melanoma by wavelet and curvelet analysis : study based on neural network classification, no. 6, pp 680–685
14. Wadhawan T et al (2011) SkinScan c : a portable library for melanoma detection on handheld devices. Departments of 1 Computer Science, 2 Electrical and Computer Engineering, and 3 Engineering Technology, University of Houston, Houston, TX 77204, USA, pp 133–136
15. Fidalgo Barata A, Celebi E, Marques J (2014) Improving dermoscopy image classification using color constancy. IEEE J Biomed Heal Info 2194(c):1–1
16. Abuzaghleh O, Barkana BD, Faezipour M, (2014) Automated skin lesion analysis based on color and shape geometry feature set for melanoma early detection and prevention. In: IEEE long island systems, applications and technology (LISAT) conference 2014
17. Jain S, Jagtap V, Pise N, Computer aided melanoma skin cancer detection using image processing. Procedia Comput Sci 48(C):736–741
18. Tan TY, Zhang L, Jiang M (2016) An intelligent decision support system for skin cancer detection from dermoscopic images. In: 2016 12th International Conference on Natural Computation, Fuzzy Systems and Knowledge Discovery (ICNC-FSKD), pp 2194–2199
19. Adjed F, Gardezi SJS, Ababsa F, Faye I, Chandra Dass S (2018) Fusion of structural and textural features for melanoma recognition. IET Comput Vis 12(2):185–195
20. Rajesh A (2018) Classification of malignant melanoma and Benign Skin Lesion by using back propagation neural network and ABCD rule. Cluster Comput 1–8
21. Oliveira RB, Marranghello N, Pereira AS, Manuel J, Tavares RS (2016) A computational approach for detecting pigmented skin lesions in macroscopic images. 61:53–63
22. Jaisakthi SM, Mirunalini P, Aravindan C (2018) Automated skin lesion segmentation of dermoscopic images using GrabCut and k-means algorithms. IET Comput Vis 12(8):1088–1095
23. Ferreira1 TMPM, JSMARSMJR (1967) Abstract—The, PH2—a dermoscopic image database for research and benchmarking. J ACM 14(4):677–682
24. H. J. Vala and P. A. Baxi, "2013-a-Review-on-Otsu-Image-Segmentation-Algorihm," vol. 2, no. 2, pp. 387–389, 2013
25. Maglogiannis I, Doukas CN, Member S (2009) Overview of advanced computer vision systems for skin lesions characterization. 13(5):721–733

Chapter 7
Fractal Analysis of Heart Dynamics During Attention Task

Mukesh Kumar, Dilbag Singh and K. K. Deepak

1 Introduction

Human attention is the ability to focus an entity in a specific location over other entities in the space [27]. Human attention functions either internally or externally [4, 6, 35]. Internally operative attention involves short-term memory [21], whereas externally operative attention is automatic and transient [20]. It was considered that these attention types realize common neuropsychological processes to improve human perception [7, 12, 13, 25], whereas few shreds of evidence reported that these attention types involve different neuropsychological processes [4, 8, 15, 20, 21, 29]. In a study, [29], observed that increased target-discrimination difficulty in Posner's spatial cueing paradigm increases cueing effect on internal attention while cueing effect reduced in an external attention task. In a subsequent fMRI study of [8], similar differences were observed for consequences of internal and external attention with a face discrimination task, and it was shown that internal attention elicited higher activation of face processing in the brain than external attention. Thus, both attention types have different consequences and may be controlled by different neuropsychological mechanisms [21].

Although numerous efforts have been made in order to quantify internally and externally operative attention, a strong and simple distinction of physiological bases of attention is not available [5, 7, 8, 15, 20, 21, 25, 26, 29, 35]. Thus, it supports the need for further investigations to understand the underlying mechanisms responsible

M. Kumar (✉) · D. Singh
Department of Instrumentation and Control Engineering, Dr B R Ambedkar National Institute of Technology, Jalandhar, Punjab, India

D. Singh
e-mail: singhd@nitj.ac.in

K. K. Deepak
Department of Physiology, All India Institute of Medical Sciences, New Delhi, India

S. Jain et al. (eds.), *Advances in Computational Intelligence Techniques*, Algorithms for Intelligent Systems,
https://doi.org/10.1007/978-981-15-2620-6_7

for neural control internal and external attention. Furthermore, heart-rate based bio feedback found to be very useful for the management of stress, depression, negative emotions, anxiety symptoms, and lower heart rate variability (HRV) associated with poor attention performance [14]. Indeed, an anomaly in the heart may lead to cognitive dysfunctions, but insight for these consequences is not yet completely understood [16, 23, 33]. Behavioral and electrophysiological markers of neuro-cardiac regulation related to human cognitive information processing may precede development accelerated training programs and pathological treatments of brain and heart disorders. The insight of this distinction will support us to design robust brain–computer interfaces, because the performance of such systems may affect by internally and externally operative attention [1, 2, 30].

Human cognition critically employs attention in the process of perception, information processing, and decision-making abilities [27]. Technological advances in nonlinear time series analysis may provide comprehensive information about heart–brain interactions [9, 10, 16].This chapter aims to investigate the heart rhythm variation during internally and externally operative attention, and a nonlinear approach based on multifractal analysis was applied [3, 11, 17]. The multifractal analysis identifies the fractal variations in heart rhythms acquired during baseline, internally and externally operative attention, and recovery sessions. The multifractal analysis involves random walk theory to compute the degree of irregularity and complexity of heart rhythm [24]. Since the evidence in the literature supports nonlinear approaches to provide insight of complex cognitive processes [16, 23, 33], therefore, it was hypothesized that fractal dimension analysis would reveal differences in internally and externally operative attention.

2 Materials and Methods

A. *Subjects*

The study includes seven males, seven females (26.35 ± 2.15 years), and healthy volunteers. All subjects were right-handed with normal or corrected to normal vision (6/6 visual acuity) and nonsmokers. Also, subjects were also instructed to refrain from alcohol, drug use, caffeinated beverages, and physical activity for four hours before the experiment [14]. The subjects have signed an informed consent form to participate in the study. People with a history of vision, cardiovascular, and neurological disorders were not considered for the study. The institute ethical committee approved the experimental protocol.

B. *Experiment Design and Procedure*

The experiment involved three stages: (1) baseline (2) attention task stimulation and (3) recovery. During the baseline, subjects were instructed to concentrate on center fixation point (+) while watching a blank gray computer screen for 6 min. Secondly, for attention task stimulation, subjects performed six sessions, each session with

120 trials (each task session includes an additional practice session of 20 trials and minimum 5 min of rest to proceed for next session) of an easy face discrimination task with alternating peripheral cue conditions to differentiate internally and externally operative attention. In the end, subjects were recorded for recovery session which involves the same procedure as follows for the baseline session. Subjects were also reminded for minimum eye blinks and not to fall asleep throughout the trials [5]. However, subjects were in a comfortable sitting posture with their chin rested in chin rest at a distance of 57 cm from the computer screen, at which 1 cm corresponds to 1° of visual angle [5]. The response keyboard and computer screen used to display stimulus presentation. Subjects performed Posner's spatial cueing paradigm (see Fig. 1), as commonly used and validated for internally and externally operative attention stimulation [4, 8, 21, 28, 29, 35]. The stimulus trial start and end frame comprehended with a fixation field that consisted of the fixation point and two target placeholder boxes for 1000 ms. The internal or external peripheral cue (one of the two rectangular placeholder boxes, one turned red for 250 ms) and face target were equally likely to appear on the right or left of the fixation cross. Internal and external cueing order was counterbalanced. The target was either of the two male faces or no face in a rectangular placeholder box (left or to right to the central fixation point for 300 ms) after a cue offset.

Target face was created from digital grayscale images of two distinct male faces with neutral expressions without non-facial features such as hairs, ear, and neck and selected from the FEI face database [34]. Subjects were instructed to answer one of the numeric numbers assigned to target faces: as "1" for first male face, "2" for second male face, and "0" for no face trial. However, the sequence and timing of

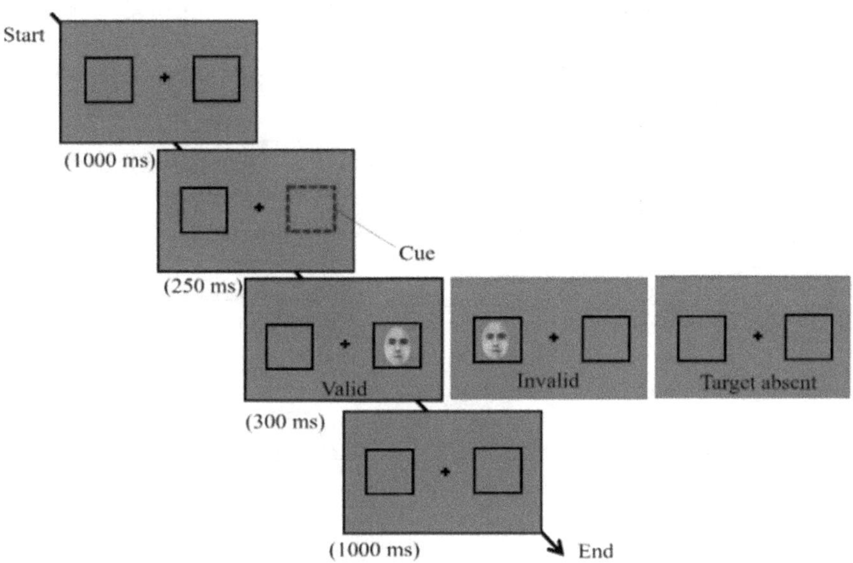

Fig. 1 Schematic diagram for spatial cueing paradigm

events in the task were identical for internally and externally operative attention stimulation except for the fact that probabilities of valid, invalid, and no-target trials, which differed to enable physiological examination of internal and external attention with identical stimulus parameters. For the internally operative attention, 70% were valid trials, 15% were invalid trials, and 15% were no face trials. For the externally operative attention, 40% were valid trials, 40% were invalid trials, and 20% were no face trials. There was a randomized order of stimuli and an even split of trials in which target appeared on the left or right side.

A. *Physiological measurements*

Subjects were made prepared for physiological measurement of ECG signal with B-Alert X10® wireless headset (Advanced Brain Monitoring, Carlsbad, CA, USA), and the surface ECG electrodes were placed on the right clavicle and lower left rib cage [1]. Then, subjects were made to relax for 15 min before proceeding to the experiment. The B-Alert X10® headset records ECG signals at a sampling frequency of 256 samples per second and 16-bit resolution, with CMRR of 105 dB and transferred in real time via Bluetooth link (2.4–2.48 GHz) to a host computer [2].

3 Data Analysis

Data are expressed as means ± standard deviation. Response time analysis was performed for both internally and externally attention task. In addition, accuracy, detection rate, and mean response times were compared for both internally and externally operative attention conditions. Subsequently, preprocessing of the ECG signal for all subjects was performed and noted as more than 93% of recordings were artifact-free-epochs. Then, ECG signals were trimmed for the first and last 30 s to remove the participant's posture adjustment artifacts.

To identify and rectify artifacts, such as invalid spikes, amplifier saturation, and undue muscle activities (EMG), the B-Alert-X10® system automatically decontaminates the artifacts, detailed artifact decontamination algorithm provided in [2]. To identify EMG artifacts, 70–128 Hz power spectral density is computed by FFT with Kaiser window ($\alpha = 6.0$) [1, 2].

Thus, trimmed ECG of five minutes was preprocessed as heart time series for multifractal detrended fluctuation analysis (MDFA), which define scaling behavior along with possible signal trends without knowing its shape and origin [22, 24]. MDFA algorithm utilizes random walk theory, for detrending local variability in a time series [18]. Prominent step in preprocessing of MDFA algorithm is to consider a nonstationary process, as self-similarity of fractal series can only apply to non-stationary processes [32]. To convert the heart time series, a summation conversion process was used [18, 19, 24]. For a heart time series $x(n)$ having length N, obtain cumulative summation $y(n)$, calculated as the normalized signal [18].

$$y(n) = \sum_{i=1}^{n} (x(i) - \langle x \rangle) \tag{1}$$

where $n = 1, \ldots, N$, $\langle x \rangle$ mean of time series, and $y(n)$ cumulative summation signal. Thus, detrended fluctuation scaling analysis considers cumulative summation signal $y(n)$ instead of $x(n)$ to qualify for the assumptions of stationarity.

To compute signal scaling profile, the resultant scaling exponent α is an evaluation of Hurst exponent H. If α is 0–1, then signal profile owns a stationary process where $= \alpha$. In case of a nonstationary process, the α is 1–2 where $H = \alpha - 1$ [17, 24]. Also, a set of window sizes s (which is equally spaced on a logarithmic scale and the number of sample or length of the signal,N) is required to define before scaling analysis [22]. Thus, split the cumulative summation signal y (n) with window set N_s, with nonoverlapping window signals of equal length. For instance, N may not be of considered size, and some fragments of the series remain left; therefore, the same scaling is repeated from the other end [17].In that way, $2N_s$ window segments will be created in total $N_s \equiv \text{int}(N/s)$. To generate detrended time series $Y_s(n)$, remove polynomial trend $y_{v,s}^m(n)$ within each segment v of time series for every window set where m is the degree of polynomial ($m = 1$: linear and $m = 2$ is quadratic), and v represents segment length ($v = 1, \ldots, 2N_s$).

To rectify unwanted noise such as the trend of the time series, subtract $y_{v,s}^m(n)$ from cumulative summation signal $Y(n)$[19]. As compared to without trend signal, the window size increases at a higher proportion for a signal with the trend. Thus, the detrending process reveals true scaling for both correlated and uncorrelated signals [24].

$$Y_s(n) = y(n) - y_{v,s}^m(n) \tag{2}$$

The alteration of the detrended time series $Y_s(n)$, in each segments $v = 1, \ldots, 2N_s$, produces the mean-square fluctuations as:

$$F_{\text{DF Am}}^2(v, s) = \frac{1}{s} \sum_{j=1}^{s} Y_s^2(n) \tag{3}$$

To obtain the qth order fluctuation function:

$$F_q(s) = \left\{ \frac{1}{2N_s} \sum_{v=1}^{2N_s} \left[F_{DFA}^2(v, s) \right]^{\frac{q}{2}} \right\}^{\frac{1}{q}} \sim s^{h(q)} \tag{4}$$

Now repeat the above mentioned procedure for different scales s. $F_q(s)$ depend on the timescale s for various values of q. The optimal value of q is between $q = -5$to 5 in biomedical signal processing applications [17, 22]. For $q = 2$, usual detrending fluctuation analysis is processed. The fluctuation function is given as $F(s) = s^{\alpha}$, where α provides fractal dimension, D related to Hurst exponent [17] as $H = 2 - D$.

Also in Eq. (4), the $h(q)$ is called the generalized Hurst exponent. For positive q value, scaling $h(q)$ reflects the scaling of the large fluctuation and strong multifractal, and for negative q value, $h(q)$ reflects the scaling of the small fluctuation and weak multifractal [17, 24]. Scaling exponent $\tau(q)$ of the measured fractal is related to $h(q)$ as

$$\tau(q) = q * h(q) \tag{5}$$

The generalized fractal dimension $D(q)$[17, 18, 24] and the singular spectrum can be derived with τ (q), as

$$D(q) = \frac{\tau(q)}{q - 1} \tag{6}$$

The grand mean of fourteen subjects for generalized fractal dimension analyzed with paired sample t-test computed for session-wise comparisons. A significance value (p) less than 0.05 (95% of confidence interval) was considered significant.

4 Results and Discussions

Mean response times for correct answers for each attention task and subject were evaluated by two-way ANOVA. For internally operative attention, the subject's response time for valid trial was 650.44 ms as compared with externally operative attention was 680.01 ms. The main effect of attention condition on response time yielded an F ratio of $F(1, 13) = 5.47$, $p < 0.05$. Moreover, the accuracy for valid trials (93.82%) was higher than invalid trials (91.09%) in internally operative attention, and for externally operative attention, the accuracy for valid trials was 95.23%, and invalid trials were 94.38%. However, detection rates during valid and invalid trials in externally operative attention condition were significantly higher than following internally operative attention condition (valid = 94.42%; invalid = 90.21% and valid = 96.41%, invalid = 93.73% of internally and externally operative attention, respectively). These results enumerate that the subjects performed better in externally operative attention than with internally operative attention. Furthermore, for multifractal analysis, the heart's rhythm acquired from fourteen subjects represented as heart time series $x(n)$. The global trend of the heart rhythm is eliminated by normalizing the DC bias to zero and by computing cumulative summation signal and represented in Fig. 2.

The obtained detrended signal profile by splitting the signal into equal sets of segments using linear least-square fit ($m = 1$) is illustrated in Fig. 3. The length of the signal taken here for analysis is 300 s with 256 sampling frequency, which results in 76,800 samples (however, only the first 10,000 samples were represented in Figs. 2 and 3). By using Eq. (3), obtain the fluctuation function and obtain the regression line of the corresponding DFA using least-square fit. The slope of the regression line

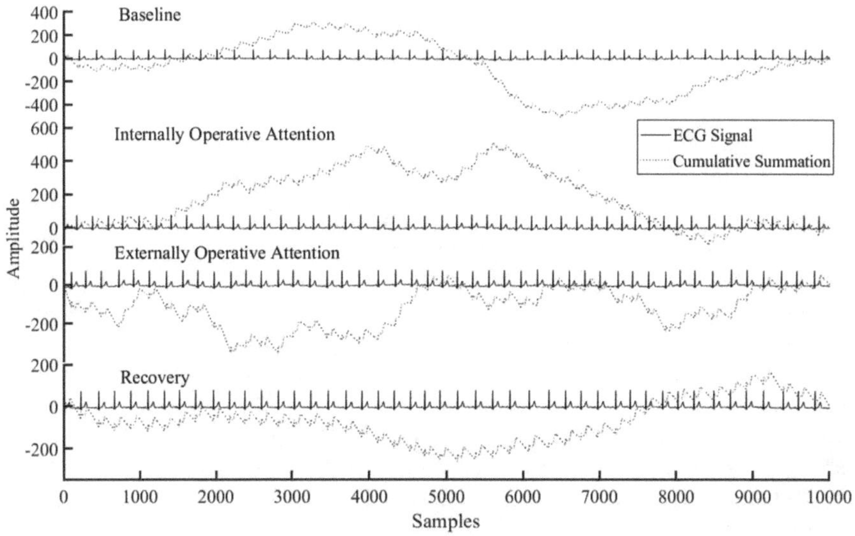

Fig. 2 Heart rhythm during baseline, internally and externally operative attention, and recovery session with its small and large fluctuations obtained in signal profile. *Note* All time series of fourteen subjects were having 76,800 samples. However, here for representation, only the first 10,000 samples were considered

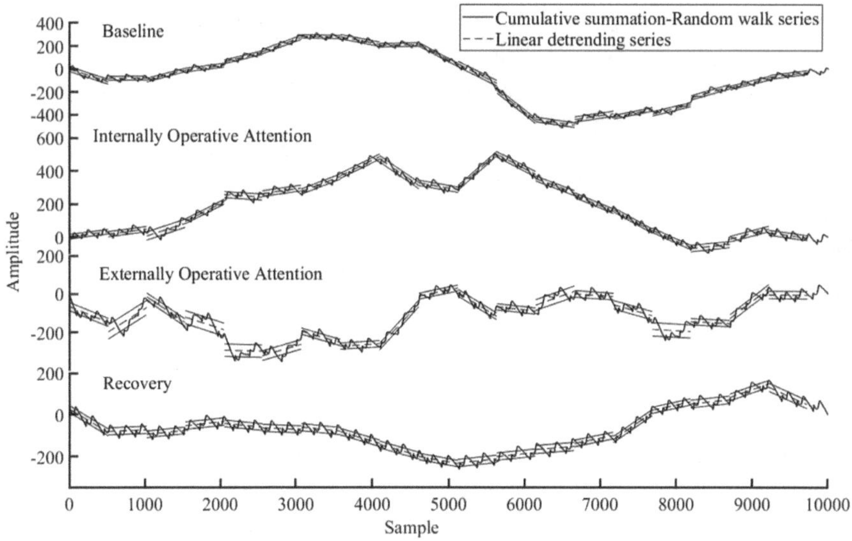

Fig. 3 Computation of cumulative summation random walk series as a detrended signal with linear least-square estimation by splitting the time series with equal segments of baseline, internally and externally attention, and recovery session. *Note* All time series of fourteen subjects had 76,800 samples. However, here for representation, only the first 10,000 samples were considered

determines the DFA exponent, α between 0 and 1, which means it can be modeled as the fractional Gaussian noise with $H = \alpha$ [18].

The physiological signals have scale invariance structures and multifractal behavior. So, multifractal spectrum width and shape differentiates the heart rhythm variations during the attention task. To study the alteration in the fractal dimension of the signal with attention conditions, heart rhythms were analyzed from baseline, internally and externally operative attention from fourteen healthy subjects, and results are enumerated in Table 1. For baseline and recovery session, fractal dimension noted with $\langle h(q)_{\text{maximum}} - h(q)_{\text{minimum}} \rangle$ as spectrum width 1.35 ± 0.21 and 1.41 ± 0.14 (mean $\pm$ standard deviation), respectively; it indicates that the heart rhythm exhibits the existence of long-range correlation and Gaussian shape as presented in Fig. 4a, d.

Moreover, for internally operative attention task, the subjects show significant variation from baseline and recovery session with a spectrum width of 1.28 ± 0.21 and Gaussian shape as presented in Fig. 4b. However, for externally operative attention task, subjects show significant variation from baseline session with a spectrum width of 1.31 ± 0.19 and Gaussian shape, as shown in Fig. 4c. For internally and externally operative attention task, the multifractal dimension analysis reveals that heart rhythm embraces fractal behavior with the reduced spectrum and no change in shape the multifractal spectrum concerning baseline and recovery session as shown in Fig. 4. For a baseline and recovery sessions, the multifractal spectrum shows a Gaussian response with increased spectrum width, while for attention task, the multifractality of spectrum is reduced (Fig. 5).

Few studies reported the use of multifractal spectrum analysis to differentiate health and deceased person; as for a healthy subject, the spectrum width is observed as pure Gaussian, and for pathological subjects, the multifractal behavior reduced [22, 24, 37]. For instance, heart rate varies with postural stimulation as when the subject is in a supine state and due to reduced gravity of human body that allows more blood flow returns to the heart, which results in decreased heart rate [22, 31]. Thus, multifractal spectrum width will be more for lying subjects than the standing subjects [22]. Similarly, the findings of this study revealed that fractal variations of heart rhythm found to quantify internally and externally operative attention sates with respect to baseline and recovery session, which may support us to design robust hybrid brain–computer interfaces [30], accelerated sports, and defense training paradigms [2].

5 Conclusion

This chapter presents nonlinear multifractal complexity analysis of the heart's activity and the level of task accuracy during internally and externally operative attention. For both attention states, the multifractal spectrum width of heart rhythm reduced as compared to baseline and recovery session. Our results revealed that subjects were more accurate during externally operative attention as compared with internally operative attention. The multifractal spectrum width of heart rhythm observed was comparable during different attentional conditions. Together, these findings indicate

Table 1 Multifractal spectrum analysis of heart rhythms during baseline, internally and externally operative attention, and recovery sessions

Task[a]	Multifractal spectrum analysis				
	$\boldsymbol{h(q)}_{\max}$	$\boldsymbol{h(q)}_{\min}$	Spectrum width	Spectrum shape[b]	Significance differences (p)
Baseline (b)	1.78 ± 0.24	0.44 ± 0.14	1.35 ± 0.21	Gaussian	$b = r,\ t = 1.10, p = 0.14$ (>0.05)
Internally operative attention (i)	1.72 ± 0.23	0.44 ± 0.11	1.28 ± 0.21	Gaussian	$i < b,\ t = 1.77, p = 0.04(< 0.05)$ $i < r,\ t = 2.25, p = 0.02(<0.05)$
Externally operative attention (e)	1.74 ± 0.21	0.43 ± 0.11	1.31 ± 0.19	Gaussian	$e = \mathrm{b}, t = 1.19, p = 0.12$ (>0.05) $e < \mathrm{r}, t = 1.19, p = 0.03(<0.05)$
Recovery (r)	1.84 ± 0.15	0.44 ± 0.12	1.41 ± 0.14	Gaussian	$i = e,\ t = 1.21, p = 0.12$ (>0.05)

[a]Data represented as mean ± standard deviation; b, i, e, and r represent the abbreviations for baseline, internally and externally operative attention, and recovery session

[b]Paired t-test comparisons were conducted for multifractal spectrum width of heart rhythm recorded during baseline, attention task, and recovery sessions with a significance level of $p < 0.05$

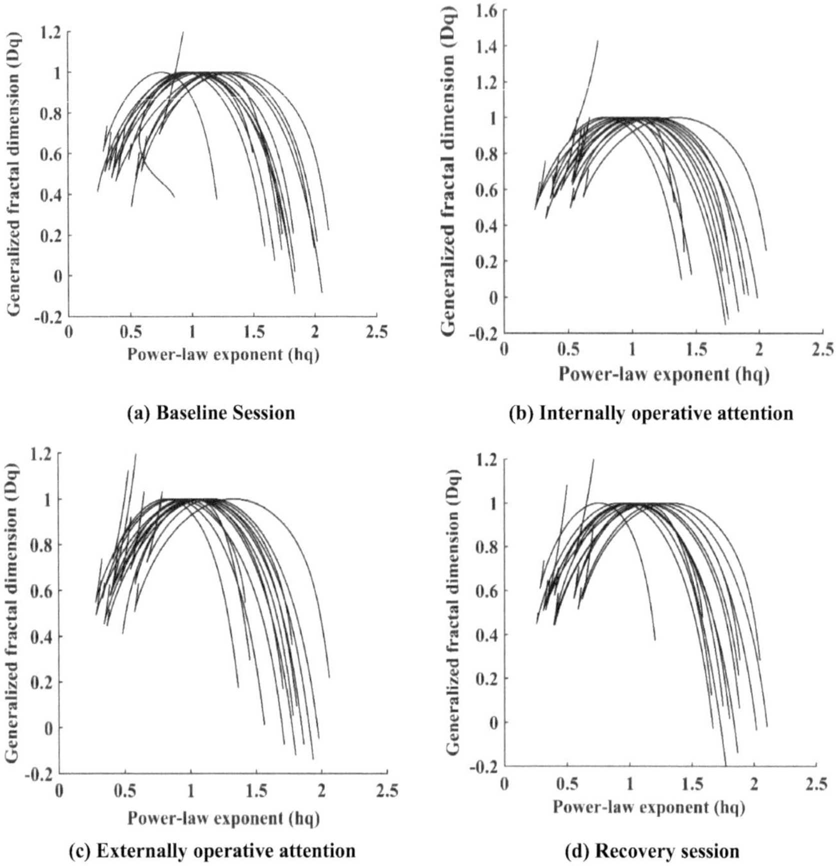

Fig. 4 Multifractal spectrum of **a** baseline, **b** internally operative attention, **c** externally operative attention, and **d** recovery session for all subjects

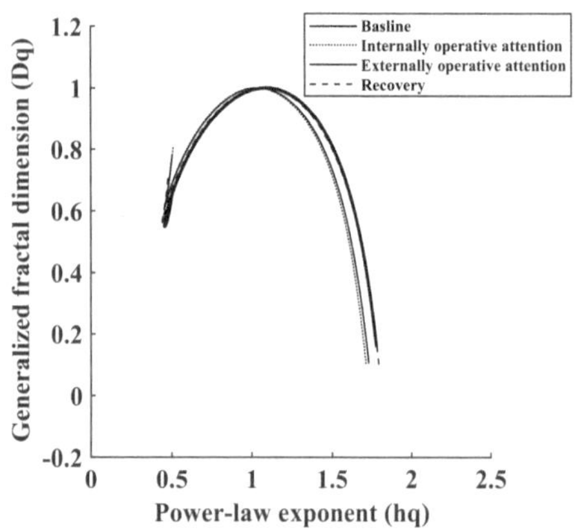

Fig. 5 Multifractal spectrum of baseline, internally and externally operative attention, and recovery session

that both internally and externally operative attention may share common autonomic resources along with different attention allocation processes. Overall, multifractal analysis reveals that the heart's rhythm possesses fractal behavior and can serve as a useful measure to understand operational activation of human attention through autonomic mobilization.

Acknowledgements Authors acknowledge support and motivation provided by faculty and staff of the department of instrumentation and control engineering, Dr. B R Ambedkar National Institute of Technology, Jalandhar, Punjab (India).

References

1. Berka C, Levendowski DJ, Cvetinovic MM, et al (2004) Real-time analysis of EEG indexes of alertness, cognition, and memory acquired with a wireless EEG headset. Int J Hum Comput Interact 17:151–170. https://doi.org/10.1207/s15327590ijhc1702_3
2. Berka C, Levendowski DJ, Lumicao MN et al (2007) EEG correlates of task engagement and mental workload in vigilance, learning, and memory tasks. Aviat Space Environ Med 78:B231–B244
3. Castiglioni P, Faini A (2019) A fast DFA algorithm for multifractal multiscale analysis of physiological time series. Front Physiol 10. https://doi.org/10.3389/fphys.2019.00115
4. Chica AB, Bartolomeo P, Lupiáñez J (2013) Two cognitive and neural systems for endogenous and exogenous spatial attention. Behav Brain Res 237:107–123. https://doi.org/10.1016/j.bbr.2012.09.027
5. Chica AB, Martín-Arévalo E, Botta F, Lupiáñez J (2014) The spatial orienting paradigm: how to design and interpret spatial attention experiments. Neurosci Biobehav Rev 40:35–51. https://doi.org/10.1016/j.neubiorev.2014.01.002
6. Chun MM, Golomb JD, Turk-Browne NB (2011) A taxonomy of external and internal attention. Annu Rev Psychol 62:73–101. https://doi.org/10.1146/annurev.psych.093008.100427
7. Corbetta M, Shulman GL (2002) Control of goal-directed and stimulus-driven attention in the brain. Nat Rev Neurosci 3:201–215. https://doi.org/10.1038/nrn755
8. Esterman M, Prinzmetal W, DeGutis J et al (2008) Voluntary and involuntary attention affect face discrimination differently. Neuropsychologia 46:1032–1040. https://doi.org/10.1016/j.neuropsychologia.2007.11.014
9. Faes L, Marinazzo D, Jurysta F, Nollo G (2015) Linear and non-linear brain-heart and brain-brain interactions during sleep. Physiol Meas 36:683–698. https://doi.org/10.1088/0967-3334/36/4/683
10. Faes L, Nollo G, Porta A (2011) Information domain approach to the investigation of cardio-vascular, cardio-pulmonary, and vasculo-pulmonary causal couplings. Front Physiol 2:1–13. https://doi.org/10.3389/fphys.2011.00080
11. Fink PW, Shultz SP, D'Hondt E et al (2018) Multifractal analysis differentiates postural sway in obese and nonobese children. Mot Control 23:262–271. https://doi.org/10.1123/mc.2016-0085
12. Gazzaniga MS, Ivry RB, Mangun GR (1998) Cognitive neuroscience: the biology of the mind, 4th edn. W. W. Norton & Company Ltd, New York, USA
13. Griffin IC, Nobre AC (2003) Orienting attention to locations in internal representations. J Cogn Neurosci 15:1176–1194. https://doi.org/10.1162/089892903322598139
14. Hansen AL, Johnsen BH, Thayer JF (2003) Vagal influence on working memory and attention. Int J Psychophysiol 48:263–274. https://doi.org/10.1016/S0167-8760(03)00073-4

15. Hopfinger JB, West VM (2006) Interactions between endogenous and exogenous attention on cortical visual processing. Neuroimage 31:774–789. https://doi.org/10.1016/j.neuroimage.2005.12.049
16. Ibáñez-Molina AJ, Iglesias-Parro S (2014) Fractal characterization of internally and externally generated conscious experiences. Brain Cogn 87:69–75. https://doi.org/10.1016/j.bandc.2014.03.002
17. Ihlen EAF (2012) Introduction to multifractal detrended fluctuation analysis in Matlab. Front Physiol 3:1–18. https://doi.org/10.3389/fphys.2012.00141
18. Kantelhardt JW, Koscielny-Bunde E, Rego HH et al (2001) Detecting long-range correlations with detrended fluctuation analysis. Phys A Stat Mech its Appl 295:441–454. https://doi.org/10.1016/S0378-4371(01)00144-3
19. Kantelhardt JW, Zschiegner SA, Koscielny-Bunde E et al (2002) Multifractal detrended fluctuation analysis of nonstationary time series. Phys Stat Mech Appl 316:87–114. https://doi.org/10.1016/S0378-4371(02)01383-3
20. Landau AN, Elwan D, Holtz S, Prinzmetal W (2012) Voluntary and involuntary attention vary as a function of impulsivity. Psychon Bull Rev 19:405–411. https://doi.org/10.3758/s13423-012-0240-z
21. Landau AN, Esterman M, Robertson LC et al (2007) Different effects of voluntary and involuntary attention on EEG activity in the gamma band. J Neurosci 27:11986–11990. https://doi.org/10.1523/JNEUROSCI.3092-07.2007
22. Mary HMC, Singh D, Deepak KK (2019) Assessment of Scale Invariance Changes in Heart Rate Signal During Postural Shift. IETE J Res 2063:1–7. https://doi.org/10.1080/03772063.2019.1604172
23. McCraty R, Zayas MA (2014) Cardiac coherence, self-regulation, autonomic stability and psychosocial well-being. Front Psychol 5:1–13. https://doi.org/10.3389/fpsyg.2014.01090
24. Mercy Cleetus HM, Singh D (2014) Multifractal application on electrocardiogram. J Med Eng Technol 38:55–61. https://doi.org/10.3109/03091902.2013.849298
25. Müller HJ, Rabbitt PMA (1989) Reflexive and voluntary orienting of visual attention: time course of activation and resistance to interruption. J Exp Psychol Hum Percept Perform 15:315–330. https://doi.org/10.1037/0096-1523.15.2.315
26. Nobre AC, Coull JT, Maquet P et al (2004) Orienting attention to locations in perceptual versus mental representations. J Cogn Neurosci 16:363–373. https://doi.org/10.1162/089892904322926700
27. Posner MI, Petersen SE (1990) The attention system of the human brain. Annu Rev Neurosci 13:25–42. https://doi.org/10.1146/annurev.ne.13.030190.000325
28. Prinzmetal W, McCool C, Park S (2005) Attention: reaction time and accuracy reveal different mechanisms. J Exp Psychol Gen 134:73–92. https://doi.org/10.1037/0096-3445.134.1.73
29. Prinzmetal W, Zvinyatskovskiy A, Gutierrez P, Dilem L (2009) Voluntary and involuntary attention have different consequences: the effect of perceptual difficulty. Q J Exp Psychol 62:352–369. https://doi.org/10.1080/17470210801954892
30. Roijendijk L, Farquhar J, Van Gerven M et al (2013) Exploring the impact of target eccentricity and task difficulty on covert visual spatial attention and its implications for brain computer interfacing. PLoS One 8. https://doi.org/10.1371/journal.pone.0080489
31. Singh A, Saini BS, Singh D (2016) An adaptive technique for multiscale approximate entropy (MAEbin) threshold (r) selection: application to heart rate variability (HRV) and systolic blood pressure variability (SBPV) under postural stress. Australas Phys Eng Sci Med 39:557–569. https://doi.org/10.1007/s13246-016-0432-3
32. Telesca L, Colangelo G, Lapenna V, Macchiato M (2004) Fluctuation dynamics in geoelectrical data: an investigation by using multifractal detrended fluctuation analysis. Phys Lett A 332:398–404. https://doi.org/10.1016/j.physleta.2004.10.011
33. Thayer JF, Hansen AL, Saus-Rose E, Johnsen BH (2009) Heart rate variability, prefrontal neural function, and cognitive performance: the neurovisceral integration perspective on self-regulation, adaptation, and health. Ann Behav Med 37:141–153. https://doi.org/10.1007/s12160-009-9101-z

34. Thomaz CE (2012) FEI face database. FEI Face DatabaseAvailable
35. Van Der Lubbe RHJ, Bundt C, Abrahamse EL (2014) Internal and external spatial attention examined with lateralized EEG power spectra. Brain Res 1583:179–192. https://doi.org/10.1016/j.brainres.2014.08.007

Chapter 8
An Efficient Algorithm for Early Diagnosis of Cervical Cancer Using Random Forest Classifier

Ajay Jangra and Anjali Deswal

1 Introduction

Generally, cervical cancer is a kind of cancer which occurs in the cervix part of the uterus. Cervix part is located in the lower part of the body and which is connected to the vagina. The major cause of cervical cancer is human papillomavirus (HPV). HPV is a sexually transmitted infection. If you have a strong immune system, then your body would not be affected by the HPV [1].

However, cervical cancer cells are slow-growing. When it developed in the body, the precancerous symptoms start occurring in the body [2]. So, it provides good opportunities for patients to prevent this disease in an early stage. As better define of detection has meant to decrease in cervical cancer. You can also reduce the risk of developing cervical cancer at an early stage by various screening tests and consult doctors. Doctors will provide the proper vaccine which will protect against HPV.

2 Causes of Cervical Cancer Cervical and Recommended Screening Tests

In this section, you will learn about the causes of cervical cancer, and other various screening tests for cervical cancer are as follows:

A. Jangra · A. Deswal (✉)
Department of Computer Science and Engineering, University Institute of Engineering and Technology (UIET), Kurukshetra University, Kurukshetra (Kuk) 136119, India

S. Jain et al. (eds.), *Advances in Computational Intelligence Techniques*, Algorithms for Intelligent Systems,
https://doi.org/10.1007/978-981-15-2620-6_8

2.1 *Causes of Cervical Cancer*

2.1.1 Genetic Factors

- **Dominant**: Cervical cancer is more dominant in females than males.
- **Family background**: In past, if a family member had cervical cancer, then a person can have the chances of cervical cancer.
- **Genetic issues**: In this genetic circumstances with natural genes are modified by producing their own offspring.

2.1.2 Lifestyle Risk Factors

- **Lack of physical activity**: Due to the lack of physical activity can increase the chances of cervical cancer to some extent.
- **Improper diet**: Proper diet plays important for our immune system and helps to increase the risk for cervical cancer.
- **Consume large quantity of alcohol**: Intake of a large amount of alcohol is dangerous to our health and also can increase the risk of cervical cancer.

2.2 *Cervical Cancer Signs and Symptoms*

Strange vaginal draining is the most outstanding indication of uterine malignancy. The cervical disease normally occurs after menopause, but due to the changes in the environmental, it is likely to occur before the menopause, cancer bleeding can be in the form of watery, blood nodes [3]. Women should not acknowledge that irregular vaginal bleeding is a piece of menopause.

Women should consult their gynaecologist if they experience any of the following symptoms:

- Unusual vaginal bleeding or discharge
- Trouble urinating
- Pelvic pain
- Pain during intercourse

2.3 *Types of Cervical Cancer*

Cervical malignant growth development can occur in any layer of the uterus, and there is no specification for cancer which consists of three layers and others layers too [4]. Types of uterus cancer are as follows:

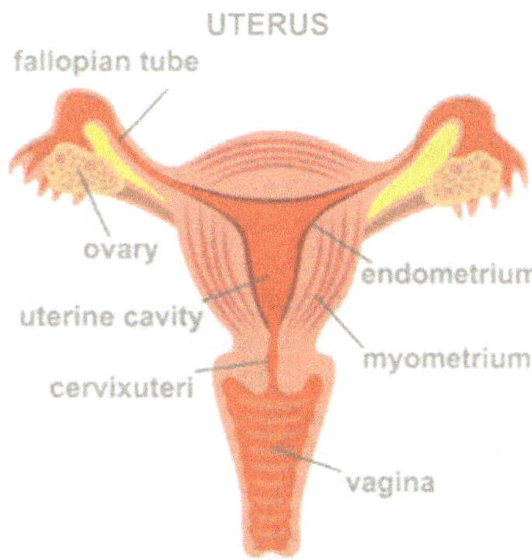

Fig. 1 Female reproductive system [5]

- **Endometrium layer**: It is the innermost layer of the uterus where implantation takes place, responsible for the menstrual cycle which is made up of connective tissues, endocrine glands and columnar epithelium.
- **Myometrium layer**: This layer is the middlemost layer of the uterus which is also known as the external layer and consists of soft muscle, responsible for the childbirth.
- **Perimetrium layer**: This layer provides support to the uterus and also known as an outer most layer of the uterus (Fig. 1).

2.4 Screening Tests

- **Pap smear test**: A Pap smear test is small which examined cervical cells under microscopic. In this test, the doctor takes the cervical cancer cells from cervix part of the uterus. Then, these cells are tested under microscopic in order to analysis cells abnormalities. This test can easily identify normal or abnormal cells which are present in the cervix part of the uterus. For younger women, Pap smear test is suggested.
- **Human papillomavirus (HPV) test**: This is a major cause of cervical cancer. In this test, cells are taken from the infection part of the uterine cervix. Generally, this test is suggested for women above 30 age and older.

3 Related Work

This section describes cancer which is present in another part of the body such as breast, lungs and skin. It is observed that there are few common characteristics amongst all these depict the fact they are in another part of the body. It will help to provide the analysis base for cervical cancer research.

Zhang et al. [6] proposed an artificial neural network (ANN) classifier performed on the small portion ($N = 174$) of the total accessible data and tested result shown more self-sufficient for the ANN classifier. The results produced by mathematical calculation are more predictive power which assures the ANN classifiers may perform better on upcoming future data. Alternatively, Krishnan and Doraiswami [7] proposed a method for extracting a new texture based on the parameters in spectra of colon images and also suggested this method can be applied for the multiple an array of abnormalities. Later on, a method was developed by Seki and Mostafa [8] which identifies protein names in the biomedical field along with the protein name boundary and method utilized a simple heuristics technique by model and combined with the word classes to avoid data problem. On the other hand, Gautam and Ismo [9] improved existing classifiers of K-nearest neighbour (KNN) distance-based and further compared the methods with real and synthetic data sets. While genetic algorithm was successfully expanded by Hill and Liverpool [10] for the multiclass classification and experiments has given efficient results with other used classifiers, support vector machine (SVM), ANN and KNN. Additionally, Chen et al. [11] provided the prediction accuracy and capability for the decision tree by performing logistic regression technique. To apply logistic and decision have shown effected result for breast cancer from the prediction models by the discovered method with a tree structure. Abou et al. [12] proposed a system which was used to identify automatic for early diagnosis of live cancer detection with an accuracy of 96.125%. This system can be used for its maximum accuracy and its computation simplicity. Similarly, Chen et al. [13] proposed a support vector machine which can rely on kernels to obtain the marker gene. Moreover, this method has achieved classification accuracy using the discovered marker genes. The future aim is to improve its performance. Hitt and Hendryx [14] proposed that researchers illustrated the ecological uprightness of currents which was significantly associated with public health in nearby regions. Although the macro-invertebrate data assessed in this research secured to the evaluated worth of maritime life. Also, the study exposed that these evaluations may also offer a recovered belief in human strength and security. While Hendryx et al. [15] proposed that the consequences of study earlier indicated on coal mining communities expressed that health differences are gathered in the mountain top mining quarter. Simply, the national goal was to repeal health inequalities which will not be accomplished by saving variations. Similarly, a methodology for mining intelligent interpretation of medical pathway data was proposed by De Weerdt et al. [16]. Mining process acknowledged as an estimable way to gain a better opinion for the exact process to operate human-centric methods like medical conditions by examining data as recorded in the information healthcare system. Sharma et al. [17] used the methodology to enhance the regularized linear discriminant report procedure. The biological importance has also been illustrated by performing a working outline. Moreover, the robustness method was presented by managing responsiveness summary and hopeful results are concerned. The distribution accuracy and ROC curve have also been addressed. A methodology was identified to decide optimal feature subset for the support vector machine (SVM) accuracy based on cardiotocography (CTG) classifier where optimal feature determination was performed by Ocak [18] using a genetic algorithm. The conclusions acknowledge by proposed scheme which was favourable to ANN and ANFIS reliable model. The process does not afford a conviction score delivered by the SVM classifier and the suggested system. Moreover, Palma et al. [19] recommended the semi-condition was build a minimum partition for drug target intercommunication points. Where density of each subset interplay edges was maximum. The outcomes recommended that semi-structured knowledge from data was decidedly able to predict new communications and improve the appearance by machine learning methods. In future work, you can search the usability to recognize unusual clusters, then compare with probabilistic soft logic (PSL)-based technique. A system was developed by Kaviarasi [20] which can predict cancer or non-cancer ineffective way and also encourage doctors to focus on the appropriate treatment for many patients. Where the decision tree algorithm and the K-means have shown results in the proposed system. Similarly, Kurniawati and Permanasari [21] obtained data sets from various data records to perform Pap smear test results from the achievement of metrics random forest tree. Where random forest tree found usable classifier amongst distinct classifiers. Meanwhile, Sutar [22] built an integrated method to obtain therapy from outcome symptom outcome, medical names, and data is obtained from a cancer forum. Further worked to improve clustering efficiency and medication recommendation. Scholar conducted 2017 [23] a survey on accuracy test for breast cancer and diabetes using WEKA tool. Where J48 gave 72.28% accurate results on breast cancer and SMO gave 76.80% accurate result on diabetes. The researchers said WEKA tool was considered the best tool for data mining classification. Kaur [24] upgraded of the ovarian disease patients with a specific end to improve the quality and able to distinguish between affected area and non-affected area. The result was obtained by using a genetic algorithm. Based on an automated detection system of effective lymph in the mammograms and for its classification was developed by conference and systems [25]. Where false positive values are greatly reduced and higher accuracy for the classification has been achieved. Similarly, Harirchi et al. [26] proposed that image features with the significant data set have been classified by using SVM classifier where the image contains 25 innocuous images and 25 malignant images by proposed algorithm efficiency has been adopted up to 94%. Hamsagayathri [27] focused on the execution of J48 and priority-based decision tree algorithm for breast cancer analysis. The result was shown on priority-based classifier with 98.5%. Additionally, Shandilya and Chandankhede [28] proposed that a survey where a new system was introduced which can overcomes some issues from the previous system also concluded that most automatic cancer prediction systems based on machine learning concepts by including classification and clustering algorithm. Likewise, Cha et al. [29] introduced an efficient algorithm which can detect oral cancer in the given particular data set which was REP tree algorithm and random tree algorithm. Hence proved to work efficiently even if data is large. Many researchers are adopting cloud computing than WEKA for detecting diseases. Moreover, Prghov et al. [30] Wisconsin Prognostic Breast cancer (WPBC) data set was used to find an efficient algorithm which can predict the cancerous or non-cancerous performance of disease by adopting data mining techniques. Where classification, C 5.0 and SVM have shown 81% precision and EM was found to be most likely for clustering algorithm with the precision of 68% also have shown classification is better predictor then clustering. Subsequently, Tamilmani and Sivakumar [31] studied various techniques for cancer discovery with different events which put advanced with an innovative scheme. A future scope is shown by combining gene-based algorithm and biological. Efficiency and least root mean square error (RMSE) of 0.0833 also recommended an algorithm which occupies minimum time to build complexities, cost and memory usage. In the meantime, Bharati et al. [32] performed methods where the data set of breast cancer was taken from UCI, and results were shown with 97.901% on K-nearest neighbours and 96.6035% on a multilayer perceptron. This paper also calculated the Kappa statistic, receiver operating characteristic (ROC) area and mean absolute error measure. ROC has shown the highest percentage with 99.9%, and multilayer perceptron have shown 98.01% of ROC area. Consequently, Singhal [33] proposed an algorithm which can diagnosed breast cancer accuracy. The results has been performed using the back-propagation algorithm and output was analysed by model, obtained seven hidden neurons has the maximum precision as compared with the other models also found lesser the mean square error more is the accuracy of the model and feed-forward backpropagation algorithm consider the best algorithm Primary. Sanjay et al. [34] proposed a new hybrid feature selection approach for the early-stage forecast of breast cancer. Data mining used this extremely effective system, and the result was received from different features choice methods which helped to diagnose cancer in the beginning stage. Moreover, Jacob et al. [35] performed a remembrance of classification and clustering algorithm where the classification algorithm was considered the genuine algorithm for prognosticating breast cancer based on the precision of the algorithm. Omar et al. [36] proposed that a lung cancer prognosis system (LCPS) which focuses on the patients suffering from the lung cancer gives the doctors and patients room to forecast and define a selected targeted treatment process to the patients' care and can improve the quality of treatments for the patients. LCPS can accept a large number of data sets and can employ several data mining algorithms to find health signs and probable outcomes. Thus, the system has decreased lung cancer death cases. Bakken and Koleck [37] proposed an application for data science which has shown successful results same as methods used by nurses and provided signifies of the various challenges and set targeted roles for primary responsibilities for using data science application in the advancement of health management.

4 Issues and Challenges

In this section, improvements for the existing uterus cancer detection system available are required. Also, most of the researchers were facing issues are as follows:

4.1 Proper Screening Tests for Cervical Cancer

The screening term shows some hope for those patients who were getting the symptoms for cervical cancer. Some of the common screening tests are needle test, Pap smear test and urine imaging. Researchers are not able to identify proper normal and abnormal cervical cancer cells. So, screening tests are required more superior ways which can be perfect and make fewer faults.

4.2 Early Diagnosis for Cervical Cancer Detection

A new technique for early diagnosis for cervical cancer detection is still required in order to save several lives. Due to the faulty system, every year several people are dying because these faulty systems are not able to detect cervical cancer cells.

4.3 Overview of Related Work Detection Techniques

Nowadays, detection techniques ways are better than the earlier time period. Four procedures are as follows pre-process, testing, segmenting and identify cervical cancer cell. For example, proper methods for identifying breast cancer are known as mammogram, and the liver infected portion can be easily viewed by the CT method. Moreover, for cervical cancer still difficulties are getting to identify exact cells.

5 Proposed Work

Aim for proposing proposed work is to save several lives and also to help those patients who are facing symptoms of cervical cancer. In the proposed work, the data set is taken from UCI machine learning repository (cervical cancer risk factors). Furthermore, the resultant of the data set is fed to WEKA to check the best classifier for proposed algorithm. The proposed algorithm is based on performance metrics such as TP matrix, confusion matrix and ROC curve.

5.1 Proposed Algorithm

In the proposed algorithm, variable n is used to store the instances of the dataset. Variable $a[i]$ is used to store the value of reproductive health risk, variable $b[i]$ is used to store the value of health risk and variable $c[i]$ is used to store the value of age. The variable reproductive health risk and health risk are used to calculate the risk factors (Fig. 2).

Initially, the input real data set is fed in order to mine features from the given datasets. Based on the input real data sets, two variables are introduced likely a reproductive health risk and disease risk. Further, based on these variables two classes have taken high risk and low risk, and finally, list of high-risk patients are calculated. List of high-risk patients is fed to random forest classifier to calculate high-risk list for the taken data set. Random forest classifier calculates the performance metrics such as TP rate, FP rate and precision, recall, F-measure and ROC curve (Fig. 3).

The proposed work result is carried out in the Eclipse software configured by using the WEKA software tool.

```
n ← dataset
mine features
for i=1 to n
        a = Age[i]
        b = Number of Sexual Partners[i]
        c = First sexual intercourse[i]
        d = Number of pregnancies[i]
        e = Hormonal Contraceptives(years)[i]
        rephealth = a>45 AND b>=5 AND c>18 AND d>=5 AND e>5
        a = Smoke(years)[i]
        b = IUD(years)[i]
        c = STDs:HPV[i]
        disrisk = a>=10 AND b>5 AND c
        if(rephealth AND disrisk)
                High Risk
        else
```

Fig. 2 Proposed algorithm

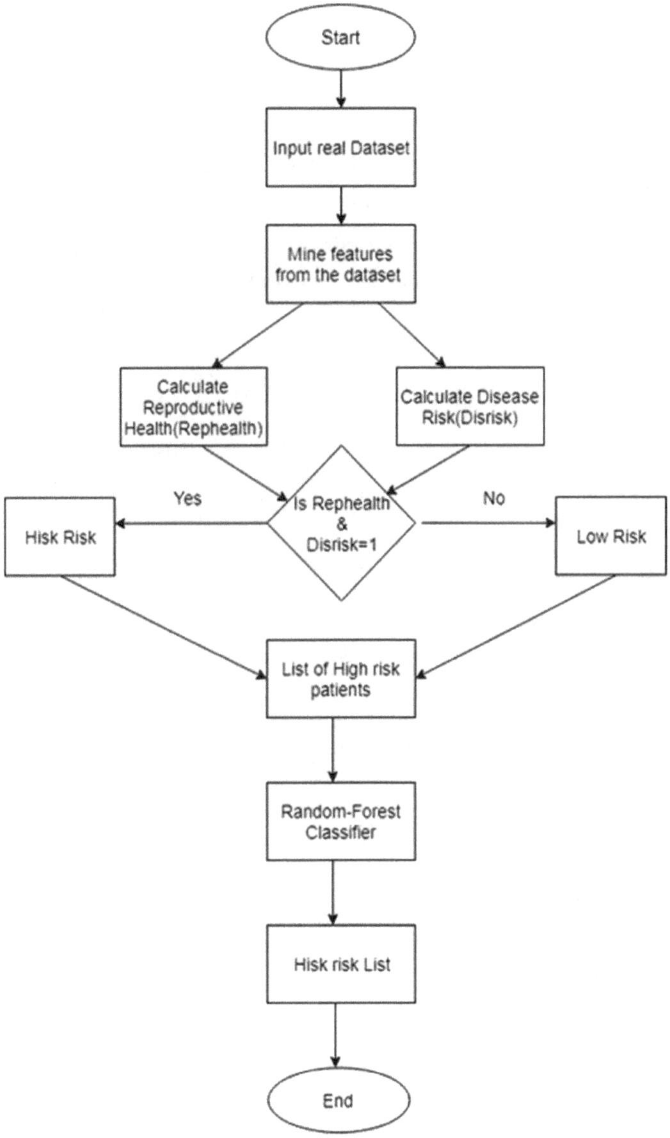

Fig. 3 Proposed framework

5.2 *Weka*

WEKA (Waikato Environment for Knowledge Analysis) is popular open-source software. It is written in Java programming language and freely available under the General Public License (GNU). WEKA was introduced at the University of Waikato,

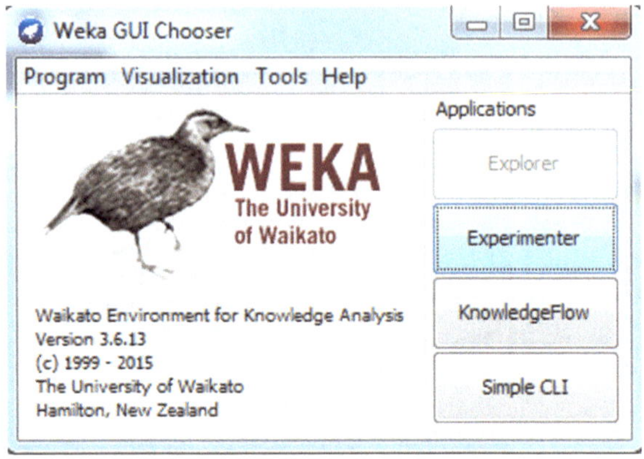

Fig. 4 WEKA tool interface, *Note* To install the WEKA tool, the latest version of Java programming language software should be installed in the system

New Zealand [38]. This tool takes "ARFF" file as an input in its explorer as shown in Fig. 4. Furthermore, after selecting the ARFF files of the data set, it performs the classification, clustering and association methods.

Interface options are usually used for the following applications:

a. Explorer: It provides an interface for exploring data set using WEKA.
b. Experimenter: It is an environment which conducts statistical tests with learning techniques
   And perform experiments.
c. Knowledge Flow: It supports incremental learning and is similar to explorer with more
   drags and drops.
d. Simple CLI: Simple command-line interface is generally used to execute WEKA's
   commands for system server.

5.3 *Explorer*

At the top, there is a line of tabs which is the only active part when the explorer is first started but before using any tab, a data set must be open in the explorer to explore the data [39] (Fig. 5).

5.4 *ARFF Data File*

An ARFF (Attribute Relation File Format) file is an ASCII text formatted file used for describing a list of the data set's where a set of attributes can be shared. Files were

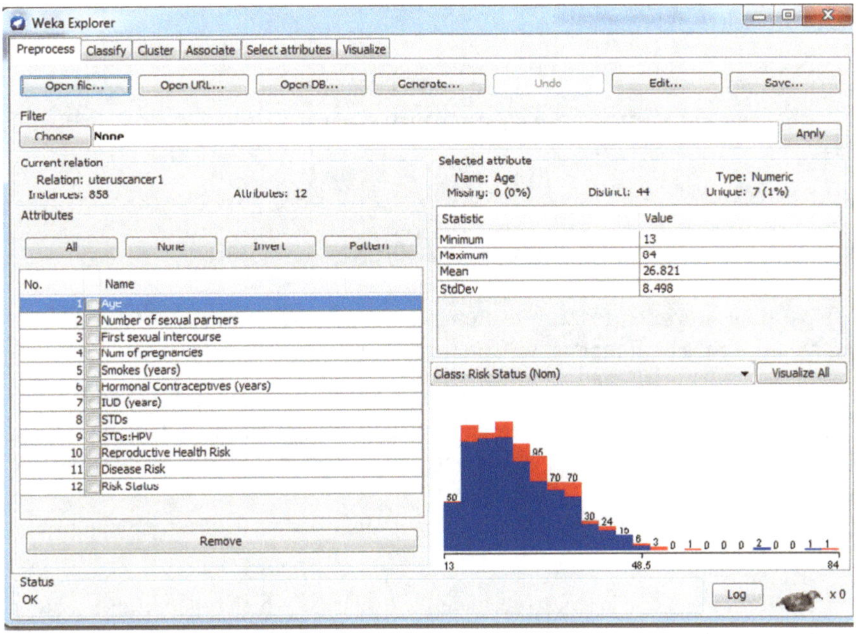

Fig. 5 WEKA explorer interface

developed for machine learning project at the University of Waikato to use WEKA. It contains two types: Header and Data. Header part contains all the relation names, list of the attributes and different types of attributes. [40].

5.5 *Eclipse (4.4) (Luna)*

Eclipse is developed from Prolog programming language. It is an open-source software tool and works as an extension where a various programming language can combine together. It is cost-effective which is used for development and constraint several programming applications. It can also use in several areas for resource allocation, scheduling, planning, etc. It has a lot of pre-installed libraries, interface to the third-party solutions and a high-level model language and control language system and also has an integrated development environment platform [41, 42].

5.6 *Performance Metrics Used*

To measure how well the designed system is performing, a social network metric called F-measure or $F1$ score is used, which can be calculated using two other metrics called precision and recall.

$$\text{precision} = \text{TPTP} + \text{FP}$$
$$\text{recall} = \text{TPTP} + \text{FN}$$

Thus, F-measure can be defined as:

$$F = 2 * \text{ precision} * \text{recall precision} + \text{recall}$$

Here, precision refers to the measurement of exactness which shows false positives and recalls term as a complete measurement which shows true positives.

- **True Positives (TP)**: Terms which were identified as correct entities.
- **False Positives (FP)**: Terms were identified as entities but should not have been.
- **True Negative (TN)**: Terms which were not identified as an entity and should not have been.
- **False Negative (FN)**: Terms that were not identified as entities but should have been.

Precision: Precision is a measurement for exactness which reflects false positive, given by the above equation if the number of false positives becomes zero; precision becomes 1 which means exactness of the proposed algorithm is perfect. Also, a large number of false positive that surpasses true positives indicates value approaching 0; poor precision.

5.7 Recall

Recall is termed as a measurement of completeness, as given by the above formula, indicates a perfect recall value if several false negatives become zero. As the number of false negatives increases, recall approaches zero.

5.8 Confusion Matrix

Confusion matrix is termed as M table which is used to define the performance for classification model using data set values, where true values are defined.

$$M = \text{TP FP}$$
$$\text{FN TN}$$

Arff Viewer: Data

Relation: uteruscancer

No.	Age Numeric	Number of sexual partners Numeric	First sexual intercourse Numeric	Num of pregnancies Numeric	Smokes Numeric	Smokes (years Numeric
1	18.0	4.0	15.0	1.0	0.0	0.
2	15.0	1.0	14.0	1.0	0.0	0.
3	34.0	1.0		1.0	0.0	0.
4	52.0	5.0	16.0	4.0	1.0	37.
5	46.0	3.0	21.0	4.0	0.0	0.
6	42.0	3.0	23.0	2.0	0.0	0.
7	51.0	3.0	17.0	6.0	1.0	34.
8	26.0	1.0	26.0	3.0	0.0	0.
9	45.0	1.0	20.0	5.0	0.0	0.
10	44.0	3.0	15.0		1.0	1.26697
11	44.0	3.0	26.0	4.0	0.0	0.
12	27.0	1.0	17.0	3.0	0.0	0.
13	45.0	4.0	14.0	6.0	0.0	0.
14	44.0	2.0	25.0	2.0	0.0	0.
15	43.0	2.0	18.0	5.0	0.0	0.
16	40.0	3.0	18.0	2.0	0.0	0.
17	41.0	4.0	21.0	3.0	0.0	0.
18	43.0	3.0	15.0	8.0	0.0	0.
19	42.0	2.0	20.0		0.0	0.
20	40.0	2.0	27.0		0.0	0.
21	43.0	2.0	18.0	4.0	0.0	0.
22	41.0	3.0	17.0	4.0	0.0	0.
23	40.0	1.0	18.0	1.0	0.0	0.
24	40.0	1.0	20.0	2.0	0.0	0.
25	40.0	3.0	15.0	3.0	0.0	0.
26	44.0	3.0	19.0	1.0	0.0	0.
27	39.0	5.0	23.0	2.0	0.0	0.
28	39.0	2.0	17.0	4.0	0.0	0.
29	37.0	3.0	24.0	1.0	1.0	3.
30	37.0	6.0	26.0	1.0	0.0	0.

Fig. 6 Data set

5.9 ROC Curve

ROC is a receiver operating statistic curve which has the ability to illustrate binary classifier in which threshold values are produced. It determines accuracy for the system. A value 1 indicates the system which is working with complete accuracy. It is plotted by using false positive rate along with the *x*-axis and true positive rate as the *y*-axis (Fig. 6).

5.10 Dataset

Various attributes used in data set:

Age: Cervical cancer mostly occurs after the premature stage because before premature stage cancer cells do not get a chance to develop. According to the other factors such as multiple intercourses, smoking and obesity increase the chance of the cervical cancer cells in the body.

A number of sexual partners: Number of sexual partners is another risk factor for cervical cancer.

First sexual intercourse: First sexual intercourse also contributes towards the risk factor for cervical cancer.

The number of pregnancies: Nowadays, having more than two or three pregnancies can cause cervical cancer in some women because of the unhealthy body.

Smoking: Smoking damages our liver system. The liver system is one important part of our body. If the liver system stops working in a human body automatically, all the functions will stop in the body which can lead to various types of cancers, and cervical cancer is also falling in the category of the other's cancer.

Hormonal Contraceptives: Hormonal contraceptives should take under the physician recommendation. Intake of access amount of contraceptives can develop cervical cancer cell.

Intrauterine Device (IUD): To some extent IUD is good, but it should take under the physician advice. Also having more than two or three times can lead to cervical cancer.

Sexually Transmitted Infections (STIs): STIs is generally caused by multiple sexual partners, and in cervical cancer HPV is the main reason for the cancer cells. HPV is caused by multiple sexual partners.

6 Results and Analysis

Various performance metrics discussed above give the following values described in Table 1 below after the recommendation model is fed to the random forest classifier.

The results described here depict various performance metrics for both the classes Low and High. As shown in Table 1, the TP rate is 1 and FP rate is zero for both the classes of diagnosed. Shown the performance for the proposed mechanism which is very accurate and also the value of *F*-measure is 1.

Table 2 reflects a confusion matrix for the proposed mechanism. Many true positive entities are 100 which show that the proposed mechanism recommends only 100 users out of the users. Thus, the proposed mechanism has optimized the performance of the early diagnosis of uterus cancer detection framework up to some extent.

Table 1 Result of performance metrics for the proposed mechanism

TP rate	FP rate	Precision	Recall	F-measure	ROC area	Classes
1	0	1	1	1	1	Low
1	0	1	1	1	1	High

Table 2 Result of confusion matrix for the proposed work

Yes	No
736	0
0	112

Fig. 7 ROC curve for the proposed work

ROC is a receiver operating statistic curve that is used to illustrate the ability of diagnosis of a binary classifier where threshold values are carried out. It determines accuracy for the system. A value 1 indicates the system is working with complete accuracy. It is plotted by using false positive rate along with the *x*-axis and true positive rate as the *y*-axis. The ROC area for the proposed system is 1 as shown in Fig. 7

The ROC area for the proposed mechanism for class Low and class High shows an area of 1 which is shown in Fig. 8 (Figs. 9, 10 and 11).

7 Conclusion

In recent times, uterus cancer has become one of the serious problems worldwide. Day by day this disease is increasing and creating serious problems in every house. Women of almost all age group are getting trapped by this disease due to lack of awareness and proper treatments. So to overcome this kind of problem in this dissertation, early

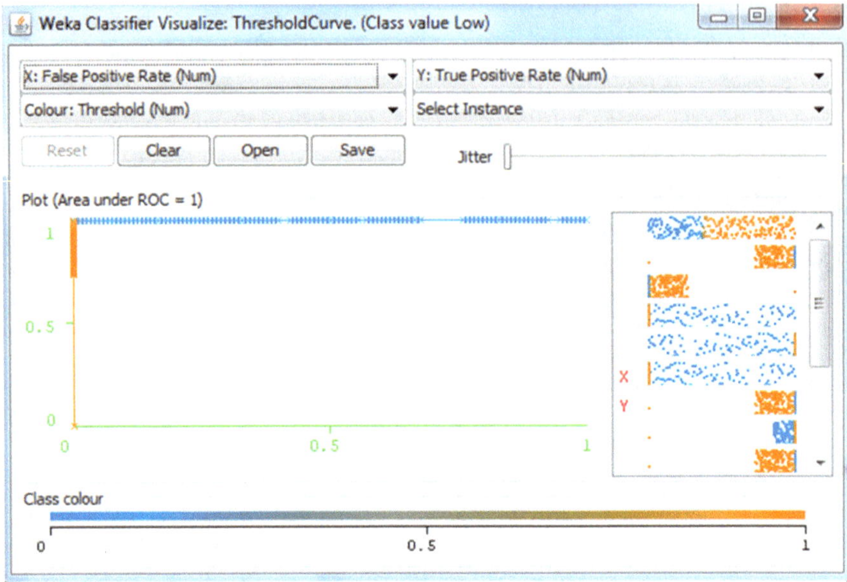

Fig. 8 ROC curve for proposed mechanism (class value Low)

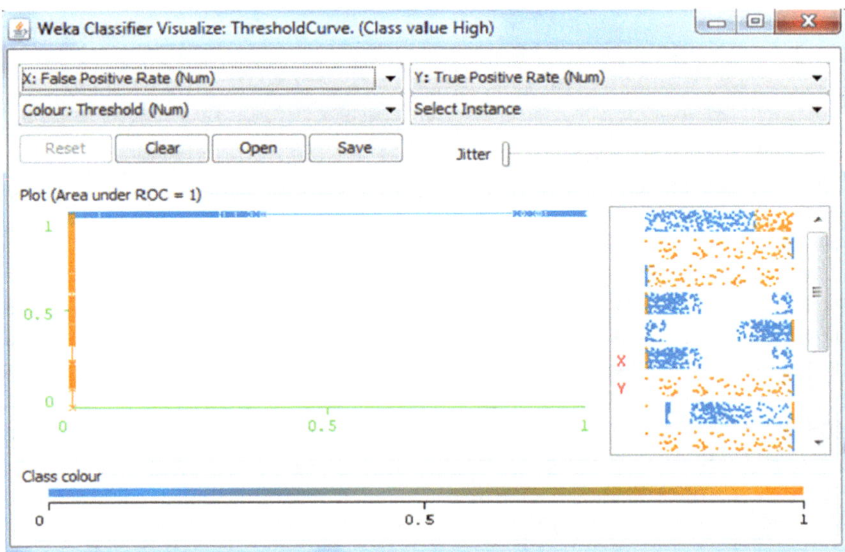

Fig. 9 ROC curve for proposed mechanism (class value High)

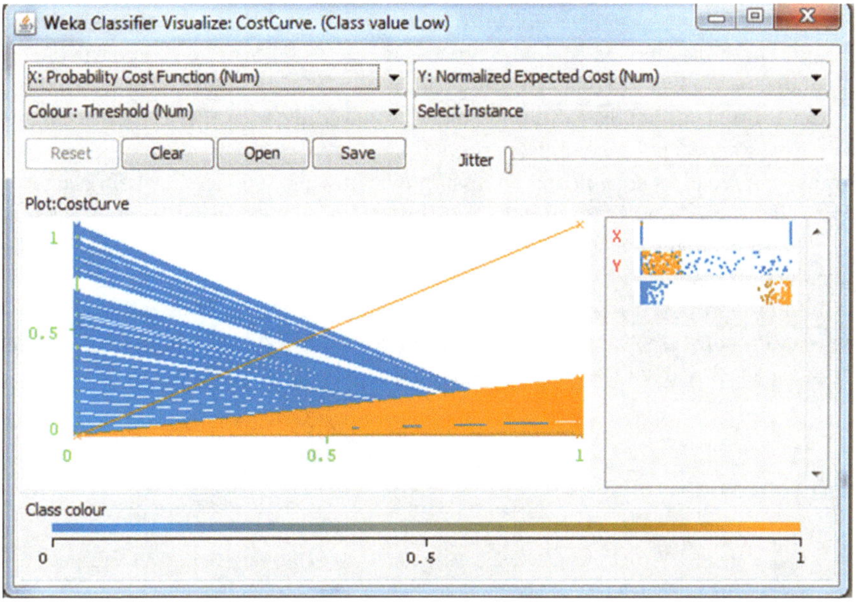

Fig. 10 Cost curve for class Low

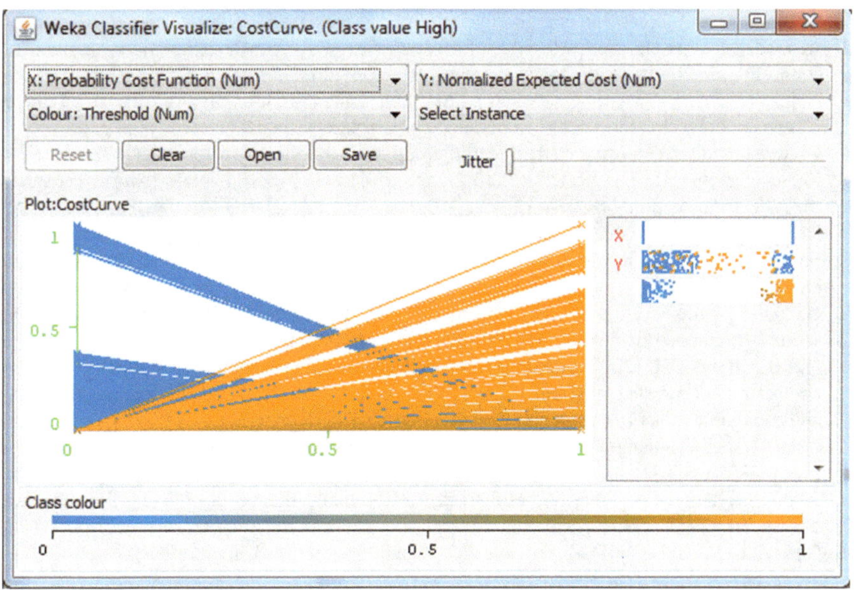

Fig. 11 Cost curve for class High

diagnosis for cervical cancer detection mechanism is proposed under supervised learning. In this method, a detection process is calculated based on some trust value calculated on the basis of a few existing parameters. Results shows that the proposed system optimally diagnosed. To implement proposed work, WEKA tool is used and to analyse the performance of proposed work different performance metrics such as *F*-measure, recall, precision and confusion matrix which are used by using random forest classifier. And proposed work was tested for all supervised learning such as decision tree, support vector machine and naïve Bayes, but random forest classifier gave the best result.

In future, proposed work can be implemented practically for unsupervised learning. Additionally, a system can be made which can identify an early diagnosis for cervical cancer and save millions of women lives.

References

1. Karssemeijer N (2010) Computer-aided detection in breast imaging: more than perception aid. In: 2010 IEEE international symposium on biomedical imaging: from nano to macro. Nico Karssemeijer Radboud, University Nijmegen Medical Centre Department of Radiology, The Netherlands, p 273
2. Vas M, Desai A (2017) Lung cancer detection system using lung CT image processing. In: International conference on computing communication control and automation, pp 1–5
3. Alsmadi O, Hussein K, Odeh M, Informatics C (2018) Towards digital cancer genetic counseling. In: 1st international conference *on* cancer care informatics, CCI, pp 188–194
4. Najeeb N, Zhang Y, Mellor C, Benson T (2015) Photonic biosensor chip for early-stage cancer diagnosis. In: 17th International conference on transparent optical networks, vol 3, pp 1–4
5. www.vectorstock.com/royalty-free-vector/structure-of-the-uterus-vector-14687872
6. Zhang Z, Zhang, H, Bast JR, RC (2000) An application of artificial neural networks in ovarian cancer early detection. pp 107–112
7. Krishnan M, Doraiswami R (2002) Based evaluation of colonoscopic images. pp 1710–1713
8. Seki K, Mostafa J (2003) A probabilistic model for identifying protein names and their name boundaries. In: IEEE computer society bioinformatics conference 2, pp 251–258
9. Gautam V, Ismo K (2004) Outlier detection using k-nearest neighbour graph. In: Proceeding of the 17th international conference on pattern recognition, pp 4–7
10. Hill B, Liverpool L (2005) Extending Genetic programming for multi-class classification by combining K-nearest neighbour. In: Zhang L, Jack LB, Nandi AK (eds) IEEE signal processing and communications group. Department of Electrical Engineering and Electronics, pp 349–352
11. Chen M, Liao H, Huang C (2006) Predicting breast tumor via mining DNA viruses with decision tree, pp 3585–3589
12. Abou S, Fakhr MW, Farag A, Mohamed A Automatic diagnosis of liver diseases from ultrasound images, pp 313–319
13. Chen Z, Li J, Wei L (2007) Maker gene identification: a multiple kernel support vector machine approach, no. 1, pp 276–279
14. Hitt NP Hendryx M (2010) Ecological integrity of streams related to human cancer mortality rates, no. 1981, pp 91–104
15. Hendryx M et al (2011) Self-reported cancer rates in two rural areas of WestVirginia with and without mountaintop coal mining. J Community Health 37:320–327
16. De Weerdt J, Caron F, Vanthienen J, Baesens B (2013) Getting a grasp on clinical pathway data: an approach based on process mining, pp 22–23

17. Sharma A, Paliwal KK, Imoto S, Miyano S (2014) A feature selection method using Improved regularized linear discriminant analysis, pp 775–786
18. Ocak H (2014) A medical decision support system based on support vector machines and the genetic algorithm for the evaluation of fetal well-being. J Med Syst 37:9913. https://doi.org/10.1007/s10916-012-9913-4
19. Palma G, Vidal M, Raschid L, Sim U (2014) Drug-target interaction prediction using semantic similarity and edge partitioning, pp 131–146
20. Kaviarasi R (2016) Recognition and anticipation of cancer and non-cancer prophecy using data mining approach. In: 2016 International conference *on* emerging trends in engineering, technology and science, pp 1–4
21. Kurniawati YE, Permanasari AE (2016) Comparative study on data mining classification methods for cervical cancer prediction using pap smear results. In: *2016 1st* International conference on biomedical engineering, pp 1–5, 2016
22. Sutar SG (2017) Intelligent data mining technique of social media for improving health care, pp 1356–1360
23. Scholar MT (2017) Using data mining classification techniques. In: 2017 international conference on intelligent sustainable systems ICISS, pp 533–538
24. Kaur B (2017) Convolutional neural network, no. Access, pp 855–859
25. Conference I, Systems I (2017) Performance analysis of mammogram, pp 1–5
26. Harirchi F, Radparvar P, Moghaddam HA, Dehghan F, Gita M (2017) Two-level algorithm for MCs detection in mammograms using diverse-adaboost-SVM, pp 269–272
27. Hamsagayathri P (2017) Priority-based decision tree classifier for breast cancer detection, pp 4–9
28. Shandilya S, Chandankhede C (2017) Survey on recent cancer classification systems for cancer diagnosis, pp 2590–2594
29. Cha J, Zea-Hernandez JA, Sin S, Graw-Panzer K, Shifteh K, Isasi CR, Wagshul ME, Moran EE, Posner J, Zimmerman ME, Arens R (2017) The effects of obstructive sleep apnea syndrome on the dentate gyrus and learning and memory in children. J Neurosci 37:3583–16. https://doi.org/10.1523/JNEUROSCI.3583-16.2017
30. Prghov F, Prghov F, Errvwlqj D (2017) Integration of data mining classification techniques and ensemble learning for predicting the type of breast cancer recurrence. Google Scholar, pp 527–530
31. Tamilmani G, Sivakumar S (2017) A survey on various data mining methods for detecting cancer cells. 34:242–245
32. Bharati S, Rahman MA, Podder P (2018) Breast cancer prediction applying different classification algorithm with comparative analysis using WEKA. https://doi.org/10.1109/ceeict.2018.8628084
33. Singhal P (2018) Artificial neural network for prediction of breast cancer. In: 2018 2nd international conference on I-SMAC (IoT SOCIAL MOBILE, ANAL CLOud) (I-SMAC) I-SMAC (IoT Social Mobile, Anal Cloud) (I-SMAC), 2018 2nd International Conference pp 464–468
34. Sanjay A, Nair HV, Murali S, Krishnaveni KS (2018) Method on real-time data. In: 2018 international conference on advances in computing, communications and informatics pp 2437–2440
35. Jacob DS, Viswan R, Manju V, Padma Suresh L, Raj S (2018) A survey on breast cancer prediction using data mining techniques. In: 2018 conference on emerging devices and smart systems, pp 256–258, 2018
36. Omar Y, Tasleem A, Pasquier M, Sagahyroon A, Lung cancer prognosis system using data mining techniques Biotech 5:361–368
37. Bakken S, Koleck TA (2019) Big data challenges from a nursing perspective" pp 3–16
38. Pazzani MJ, Billsus D (2007) Content-based recommendation systems. Adapt Web 4321(1):325–341
39. Zhang JD, Chow CY, Li Y (2015) iGeoRec: a personalized and efficient geographical location recommendation framework. IEEE Trans Serv Comput 8(5):701–714

40. Hall M, Frank E, Holmes G, Pfahringer B, Reutemann P, Witten IH (2009) The WEKA data mining software: an update. ACM SIGKDD Explor Newslett 11(1):10–18
41. Aggoun A, Chan D, Dufresne P, Falvey E, Grant H (2017) Eclipse, user manual, Mumbai. Cisco Systems Inc, India
42. Singh N, Lakhina U, Elamvazuthi I, Jangra A, Singh AK (2018) Biomedical data privacy enhancement architecture based on multi-keyword search technique. In: 2018 international conference on intelligent and advanced system (ICIAS), Kuala Lumpur Malaysia

Chapter 9
Study on Distance-Based Monarch Butterfly Oriented Deep Belief Network for Diabetic Retinopathy and Its Parameters

S. Shafiulla Basha and K. Venkata Ramanaiah

Nomenclature

DR	Diabetic retinopathy
CAD	Computer-aided diagnosis
SVM	Support vector machine
MBO	Monarch butterfly optimization
D-MBO	Distance-based MBO
PCA	Principal component analysis
2D	Two-dimensional
FM	Frequency modulation
AM	Amplitude modulation
KNN	K-nearest neighbors
GMM	Gaussian mixture model classifier
ANN	Artificial neural network
MF	Median filter
NPDR	Non-proliferative diabetic retinopathy
MAs	Microaneurysms
ANN	Artificial neural network
ROI	Region of interest
DM	Diabetes mellitus
PSO	Particle swarm optimization
NPV	Negative predictive value
GA	Genetic algorithm
CDF	Cumulative distribution function
RBM	Restricted Boltzmann machines
MLP	Multi-layer perception

S. Shafiulla Basha (✉) · K. Venkata Ramanaiah
Y.S.R. Engineering College of Yogi Vemana University, Proddatur, Andhra Pradesh 516360, India

S. Jain et al. (eds.), *Advances in Computational Intelligence Techniques*,
Algorithms for Intelligent Systems,
https://doi.org/10.1007/978-981-15-2620-6_9

GMM	Gaussian mixture model
CD	Contrastive divergence
MCC	Mathews correlation coefficient
MBO	Monarch butterfly optimization
CLAHE	Contrast limited AHE
FCM	Fuzzy C-mean
FPR	False positive rate
DBN	Deep belief network
FDR	False discovery rate
FNR	False negative rate

1 Introduction

Nowadays, DR is regarded as an important issue experienced by diabetic patients. Further, DR is the widespread and rigorous eye disease that often paves the way for vision loss and blindness for new age population. Premature detection of the disease might avoid the loss of vision in humans. DR [1–3] in premature stage could be recognized by certain characteristics such as blood vessel leakiness, irregularities, and so on. In general, DR emerges while diabetes affects the blood vessels. At the initial stage, the arteries in retina start to leak and small hemorrhages are formed. The vessels that are leaking from residuals of lipoproteins create a blur vision. A subsequent problem is the development of weak blood vessels that rupture and seeps out blood into the eye; as a result, the retina could not project images to the brain, which leads to blindness.

Diagnosis of DR in the earlier stage [4–6] is considered as the most important dilemma that requires monitoring as well as screening as early as possible. Furthermore, appropriate screening leads to premature diagnosis, timely treatment, which also diminishes the chances of total blindness. Nowadays, CAD system [7, 8] is introduced so as to diagnose DR by extracting optic disc by taking into account of aforementioned issues in the prevailing systems. In addition, the CAD systems [9–11] assist the clinicians in their process of decision-making which has increased, thus facilitating quicker and more precised diagnostic decisions with less inconsistency. Moreover, CAD [12, 13] offers a fundamental role in dealing with the elimination of preventable blindness by automated recognition of retinal pathologies [14, 15]. Furthermore, it is known that 80% of diabetics will have no indications of DR. As a result, with automated screening, only those patients with expected pathology require to be consigned. The automated detection requires more time, and in addition, it includes several individuals. Thus, in the implemented scheme, steps are taken to enhance the shortcomings of the conventional approaches.

2 Notable Methods Used in Diabetic Retinopathy Detection

2.1 Related Works

2.1.1 Pre-processing

Gabor filter: "In image processing, a Gabor filter, named after Dennis Gabor, is a linear filter used for texture analysis that basically analyzes whether there is any specific frequency content in the image in specific directions: localized region around the point or region of analysis. Frequency and orientation representations of Gabor filters are claimed by many contemporary vision scientists to be similar to those of the human visual system" [16].

2.2 Contrast Enhancement

Contrast limited adaptive histogram equalization: "CLAHE is a variant of adaptive histogram equalization in which the contrast amplification is limited, so as to reduce this problem of noise amplification. In CLAHE, the contrast amplification in the vicinity of a given pixel value is given by the slope of the transformation function" [17].

2.2.1 Segmentation

Morphological reconstruction by dilation: Morphological reconstruction can be thought of conceptually as repeated dilations of an image, called the marker image, until the contour of the marker image fits under a second image, called the mask image. In morphological reconstruction, the peaks in the marker image "spread out" or dilate [16].

Double Threshold: "This operator applies a double threshold filter to multibeam variables for the purpose of removing data (e.g., noise) within a specified range. It applies two thresholds to the data as follows: values that are equal to or above the upper threshold are unchanged." This technique was adopted in [18].

Feature Extraction: "Feature extraction may be based on ROI for retinal images by a feature vector. Because features help in differentiating EX and non-EX regions, feature set is selected for each candidate lesion using four main descriptors: area, perimeter, circularity, and diameter." This method of feature extraction was adopted in [16].

Amplitude Modulation–Frequency Modulation (AM–FM) processing: "AM–FM models describe an image in terms of highly localized 2D FM and AM. These models are remarkably powerful for characterizing and processing textured images.

The FM part of the model comprises instantaneous frequency vectors that capture the local texture orientation and granularity, whereas the AM part captures the local texture contrast" [19].

2.3 Dimension Reduction of Features

Principal component analysis (PCA): "PCA is a statistical procedure that uses an orthogonal transformation to convert a set of observations of possibly correlated variables (entities each of which takes on various numerical values) into a set of values of linearly uncorrelated variables called principal components." The dimension reduction using PCA was adopted in [19].

2.4 Classification

Probabilistic classifier: "In machine learning, a probabilistic classifier is a classifier that is able to predict, given an observation of an input, a probability distribution over a set of classes, rather than only outputting the most likely class that the observation should belong to." It was adopted in [16].

Geometrical classifier: "Geometrical classifier interval classification is a type of classification scheme for classifying a range of values based on a geometric progression. In this classification scheme, class breaks are based on class intervals that have a geometrical series" [16].

KNN classifier: "The KNN algorithm is a simple, supervised machine learning algorithm that can be used to solve both classification and regression problems. It is easy to implement and understand but has a major drawback of becoming significantly slows as the size of that data in use grows" [16].

Tree-based classification: "A classification tree is very similar to the regression tree, except the output is qualitative rather than quantitative. Recall that in a regression setting, the predicted response for observation is given by the mean response of training observations that belong to the same terminal node." This mode of classification was adopted in [16].

GMM classifier: "The GMM is basic but useful supervised learning classification algorithm to classify a wide variety of N-dimensional signals. The GMM algorithm is a good algorithm to use for the classification of static postures and non-temporal pattern recognition" [20].

SVM classifier: "An SVM is a discriminative classifier formally defined by a separating hyperplane. In other words, given labeled training data (supervised learning), the algorithm outputs an optimal hyperplane which categorizes new examples" [20].

K-means clustering: "K-means clustering is one of the simplest and popular unsupervised machine learning algorithms. In other words, the K-means algorithm identifies *k* number of centroids and then allocates every data point to the nearest cluster, while keeping the centroids as small as possible" [19].

MLP: "A MLP is a class of feed-forward ANN. The term MLP is used ambiguously, sometimes loosely refer to any feed-forward ANN, and sometimes strictly refer to networks composed of multiple layers of perceptrons (with threshold activation). MLP is sometimes colloquially referred to as "vanilla" neural networks, especially when they have a single hidden layer." This method of classification was adopted in [17].

3 A New Diabetic Retinopathy Diagnosis Model

3.1 Diabetic Retinopathy Architecture

A new DR detection model was modeled in [21], and its diagrammatic representation is demonstrated in Fig. 1. The model introduced in [21] included four phases such as pre-processing, blood vessel segmentation, feature extraction, and classification. There, pre-processing was carried out at the initial process, which enhances the image information and features. The pre-processing was carried out by models such as median filtering and CLAHE. At a particular value, the amplification was limited by CLAHE through grasping the histogram. Likewise, median filtering was a nonlinear digital filtering framework, which was exploited for eliminating noise in a given image. In addition, during the segmentation of blood vessels, the image was subdivided into several pixels for simplifying the analysis. Accordingly, FCM thresholding was exploited for rough clustering, which was further utilized for attaining the optimal threshold values. In addition, local features as well as morphological transformation-oriented features were also utilized. The local features that included the intensity of every pixel, the first-order derivatives of 2D Gaussian filtering, the second-order derivatives of Gaussian filtering and Gaussian filtering in the x-direction, and the second-order derivative of Gaussian filtering in the y-direction were mined. Further, the morphological transformation-oriented features such as bottom-hat transformation like multi-scale and multi-orientation linear structuring element were also extracted. In addition, the features were classified, by which the given image could be detected whether it was affected by DR or not. In addition, for classification purposes, DBN classifier was exploited, where the hidden neurons were optimized by means of modified MBO termed as D-MBO algorithm for increasing the accuracy rate. Thus, the D-MBO-based DR diagnosis model was proved to be more advanced in [21].

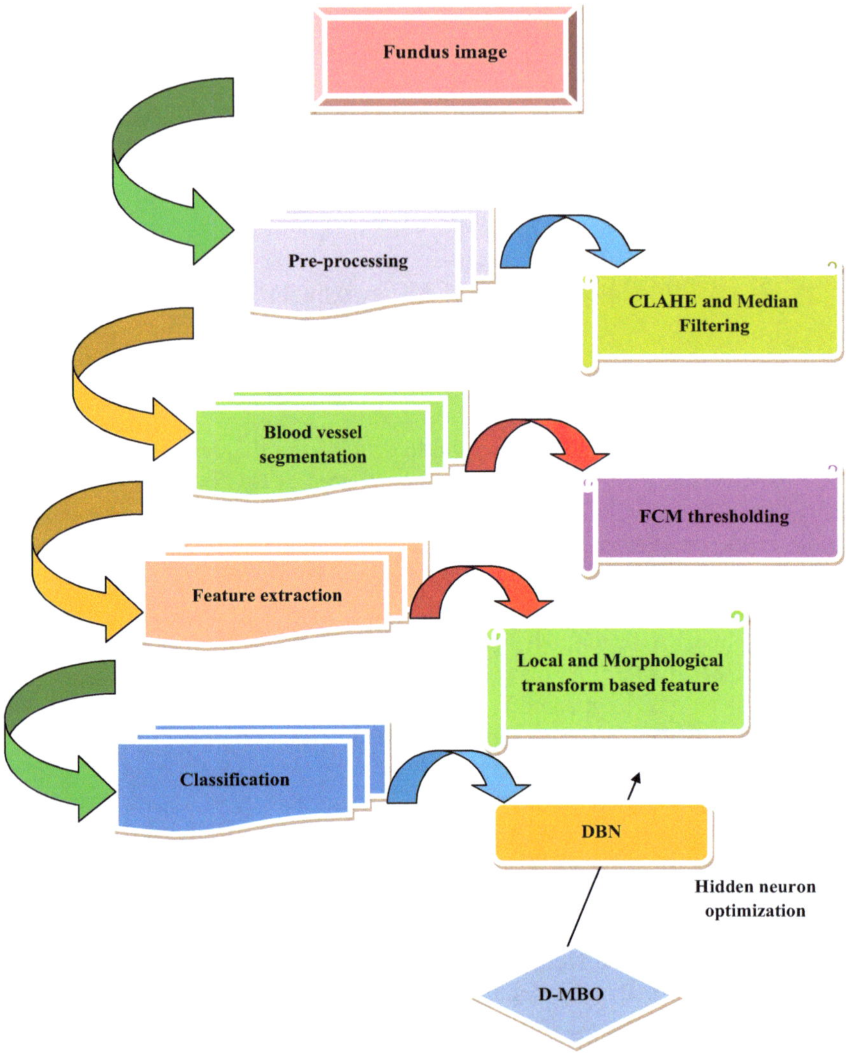

Fig. 1 Block diagram of the DR detection model

3.2 *Pre-processing*

In this study, pre-processing is carried out at the initial stage, where the input image I_{in} is pre-processed using CLAHE and median filter.

CLAHE [22]: CLAHE includes five varied stages for dividing I_{in}. Initially, the image is split into rectangular sections with equivalent size. At every division, histogram is tuned depending on the production, redistribution, and clipping. In this

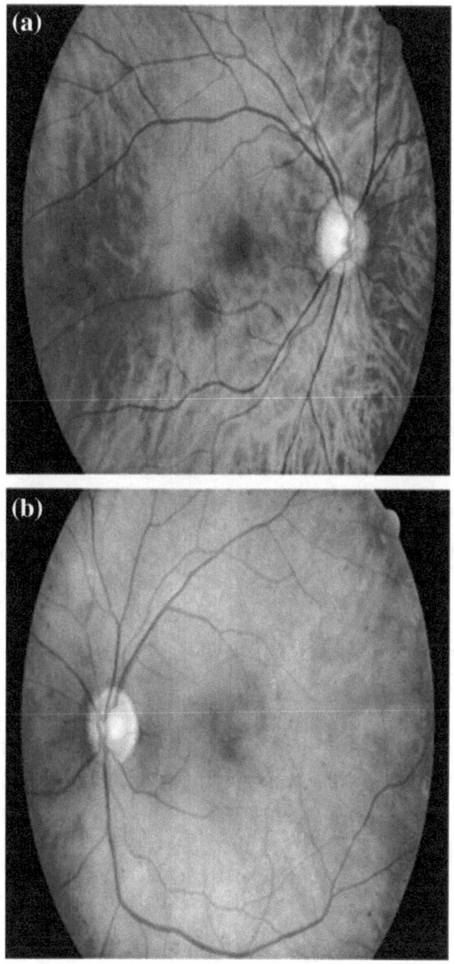

Fig. 2 Representation of CLAHE output under **a** Normal **b** Abnormal image

study, CDF is exploited for attaining the mapping function of the clipped histogram. Here, the image noise is eliminated via bilinear interpolation. Subsequently, the assessment of clip point is carried out for minimizing the highest histogram value of all divisions that is portrayed as per Eq. (1), in which η refers to the clip point, P_{i} points out the number of pixels in every section, d_{r} symbolizes the dynamic range in the corresponding block, σ refers to the clip factor, and M_{s} points out the higher slope.

$$\eta = \frac{P_{\mathrm{i}}}{d_{\mathrm{r}}}\left(1 + \frac{\sigma}{100}M_{\mathrm{s}}\right) \tag{1}$$

Therefore, I_{clahe} is the resulting pre-processed image attained by CLAHE. Figure 2 shows the sample representation of the two types (normal and abnormal) of image outputs attained from CLAHE model.

Median Filter [23]: MF is a nonlinear filter, which is deployed for substituting the noisy pixels by the adjacent pixels' median values that are sorted depending on the gray level of the image. The median filter deployed for I_{clahe} is portrayed as per Eq. (2), in which $I_{\mathrm{clahe}}(x, y)$ and $g(x, y)$ refers to the original and output image, correspondingly, and H refers to the 2D mask.

$$g(x, y) = \mathrm{med}\{I_{\mathrm{clahe}}(x - j, y - k) j, k \in H\} \tag{2}$$

Finally, the image attained after performing the median filtering process is denoted by I_{median}. Figure 3 shows the sample representation of the two types (normal and abnormal) of image outputs attained from median filtering model.

3.3 *Blood Vessel Segmentation*

The pre-processed image is provided as input for the segmentation process via FCM clustering framework [24]. Here, M elements are split into X classes, and the entire image pixels are denoted by M, which is specified as $M = M_{\mathrm{a}} \times M_{\mathrm{b}}$ and $X = 3$ denoting three-class FCM clustering. The approximation of weighted similarity is deployed in FCM technique among pixels of all cluster centers and image in an iterative optimization together with objective function. The minimized objective function is portrayed as specified in Eq. (3), in which $w_n \in W$, $W = \{w_1, w_2, w_3, \ldots, w_M\}$ points out the computed data, and $h_l \in H$ indicates $H = \{h_1, h_2, h_3, \ldots, h_X\}$ as the cluster center.

$$\mathrm{Uf} = \sum_{l=1}^{X} \sum_{n=1}^{M} (e_{\mathrm{ln}})^g \|w_n - h_l\|^2 \tag{3}$$

The standard $\| * \|$ is exploited for indicating the equivalence among all measured data and cluster centers; the real integer is denoted by $g \in [1, \infty]$ beginning from one for indicating a weighting exponent. Depending on the computation, $g = 2$ is considered as the optimal value of the fuzzification conditions as the best value for g lies among [1.5, 2.5]. The fuzzy partition matrix $E = [e_{\mathrm{ln}}]$ for $l = 1, 2, 3, \ldots, X$ and $n = 1, 2, 3, \ldots, M$ where e_{ln} refers to the degree of membership of nth pixel to lth cluster. A membership function has to solve the parameters portrayed in Eq. (4).

$$\sum_{l=1}^{X} e_{\mathrm{ln}} = 1 \quad \mathrm{for} n = 1, 2, 3, \ldots, M; \; 0 < \sum_{n=1}^{M} e_{\mathrm{ln}} < M$$
$$\mathrm{for} l = 1, 2, 3, \ldots, X 0 \le e_{\mathrm{ln}} \le 1 \tag{4}$$

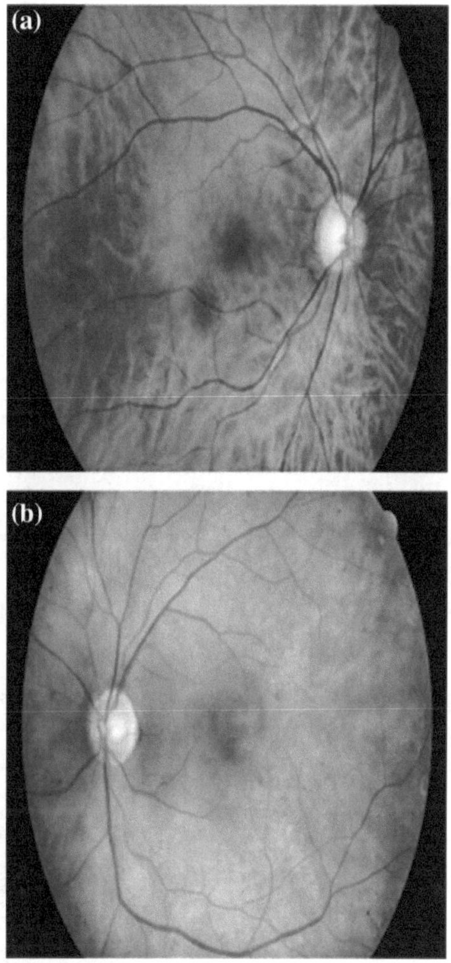

Fig. 3 Representation of median filtering output under **a** Normal **b** Abnormal image

For discovering the best fuzzy c-partition, the cluster center has to be assessed and the fuzzy partition matrix $E = [e_{\text{ln}}]$ should be developed continually. In this study, the FCM framework reduces the objective function by continual updating of membership functions and cluster center as portrayed in Eqs. (5) and (6).

$$h_l = \left(\sum_{n=1}^{M} (e_{\text{ln}})^g \right) \Big/ \left(w_n \sum_{n=1}^{M} (e_{\text{ln}})^g \right) \tag{5}$$

$$e_{\text{ln}} = 1 \Big/ \left(\sum_{q=1}^{X} \left(\frac{\|h_l - e_n\|}{\|h_l - e_q\|} \right)^{\frac{2}{g-1}} \right) \tag{6}$$

Eventually, the membership value is required to be high and based on that all the pixels are fixed to the cluster after the execution of FCM clustering. The average value is computed by the exploitation of minimum of three clusters and maximum of two clusters or minimum of two clusters and maximum of one cluster, which is then utilized for evaluating the threshold value by intensity distribution attained from the histogram of the image. By exploiting this framework, different images are captivated from diverse states and are subjected to achieving the finest threshold result. Therefore, the outcome of blood vessel segmented image attained from FCM process is denoted by I_{fcm}. Figure 4 shows the sample representation of the two types (normal and abnormal) of image outputs attained from segmentation model.

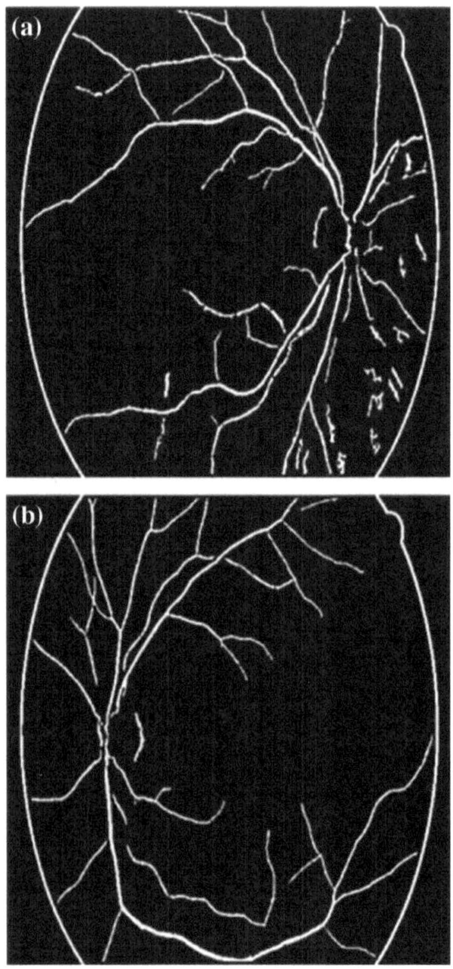

Fig. 4 Representation of median filtering output under **a** Normal **b** Abnormal image

3.4 Feature Extraction

Features such as local features and morphological transformation-oriented features are extracted in this process.

Local Features [25]: Usually, **the** intensity level of pixels is darker in the fundus images. Therefore, from the green channel image, the initial features are extracted and the leftover features are extracted from the boundaries of vessel region. Accordingly, the filtering of image handles four scales together with $\left(\sqrt{2}, 2, 2\sqrt{2}, 4\right)$. The area of the vessel is emphasized by means of 2D Gaussian filter. The 2D Gaussian kernel is computed as per Eq. (7).

$$G_k(x, y) = \frac{1}{2\pi\sigma^2}\exp\left(-\frac{x^2 - y^2}{2\sigma^2}\right) \tag{7}$$

Both Eqs. (8) and (9) define the first-order derivatives of the kernel.

$$\frac{\partial G_k}{\partial x} = \left(-\frac{1}{2\pi\sigma^4}\right)x\exp\left(-\frac{x^2 - y^2}{2\sigma^2}\right) \tag{8}$$

$$\frac{\partial G_k}{\partial y} = \left(-\frac{1}{2\pi\sigma^4}\right)y\exp\left(-\frac{x^2 - y^2}{2\sigma^2}\right) \tag{9}$$

The second-order derivatives of kernel are defined by Eqs. (10), (11), and (12).

$$\frac{\partial^2 G_k}{\partial x^2} = \left(-\frac{1}{2\pi\sigma^4}\right)\left(1 - \frac{x^2}{\sigma^2}\right)\exp\left(-\frac{x^2 - y^2}{2\sigma^2}\right) \tag{10}$$

$$\frac{\partial^2 G_k}{\partial y^2} = \left(-\frac{1}{2\pi\sigma^4}\right)\left(1 - \frac{y^2}{\sigma^2}\right)\exp\left(-\frac{x^2 - y^2}{2\sigma^2}\right) \tag{11}$$

$$\frac{\partial^2 G_k}{\partial x\partial y} = \left(-\frac{1}{2\pi\sigma^4}\right)\left(1 - \frac{xy}{\sigma^2}\right)\exp\left(-\frac{x^2 - y^2}{2\sigma^2}\right) \tag{12}$$

The Gaussian kernel in i direction $h(i)$ and j direction $h(j)$ for the second-order derivative is portrayed in Eqs. (13) and (14).

$$h(x) = \frac{1}{2\pi\sigma_1^2}\exp\left(-\frac{x^2}{2\sigma_1^2}\right) \tag{13}$$

$$h(y) = \frac{1}{2\pi\sigma_2^4}\exp\left(-\frac{y^2}{2\sigma_2^2}\right)\left(\frac{y^2}{\sigma_2^2} - 1\right) \tag{14}$$

where σ_1 is the standard deviation of Gaussian, which denotes 3 and standard deviation $\sigma_2 \in \left(\sqrt{2}, 2, 2\sqrt{2}, 4\right)$ refers to the scale parameter for the matched filter. At last, the mined local features are indicated by F_{loc}.

Morphological Transformation [25]: Similarly, I_{fcm} is further processed for extracting morphological transformation-oriented features. Top-hat transformation is processed for light pixels in the unilluminated background, and bottom-hat transformation is the converse process. The vessel region is unilluminated, and this is portrayed by bottom-hat transformation as per Eqs. (15) and (16).

$$S_i^\theta = I_{\text{fcm}} \cdot K_i^\theta - I_{\text{fcm}} \tag{15}$$

$$S_i = \sum_{\theta \in M_A} S_i^\theta \tag{16}$$

In Eq. (15), I_{fcm} refers to the input image, $\cdot$ points out the closing function, K_i symbolizes the linear structuring item $i \in \{3, 7, 11, 15, 19, 23\}$, i refers to the length of the item, S_i symbolizes the bottom-hat transformation, and the angular rotation is indicated by θ, $\theta \in M_{\text{A}}$, $M_{\text{A}} = \left\{x \middle| 0 \le x \le \pi, x \bmod \left(\pi / 12\right) = 0\right\}$.

The features are extracted by multi-scale and multi-orientation linear structuring components in bottom-hat transformation. At last, F_{mor} is achieved in this phase. The extracted feature is the mixture of both morphological transformation features and local features. Therefore, F_{final} is computed as $F_{\text{final}} = F_{\text{loc}} + F_{\text{mor}}$.

3.5 *Deep Belief Network*

The last phase in this study is the classification process, in which the extracted features F_{final} are provided as input to the optimized DBN framework. DBN [26] is a classifier with numerous layers. In each layer, the visible neurons particularly form the input layer and the hidden neurons form the output layer. Actually, hidden and visible neurons are associated with one another. In addition, there is a distinctive characteristic of the DBN framework, i.e., there is no association among two hidden neurons, and likewise, there is no association among two visible neurons. On the other hand, in Boltzmann network, the output portrayed by the stochastic neuron is probabilistic in nature. Equation (17) specifies the relevant output $\overline{PO}$, which is denoted by binary values.

$$\overline{PO} = \begin{Bmatrix} 1 \text{ with } 1 - \bar{P}_q(\zeta) \\ 0 \quad \text{with } \bar{P}_q(\zeta) \end{Bmatrix} \tag{17}$$

The sigmoid-shaped function exploited for indicating the probability function is given by Eq. (18), in which t^p refers to the pseudo-temperature and usually t^p will be greater than zero. When the value of t^p reaches zero, it is defined as per Eq. (19).

$$\bar{P}_q(\zeta) = \frac{1}{1 + e^{\frac{-\zeta}{t^p}}} \tag{18}$$

$$\lim_{t^P \to 0^+} \bar{P}_q(\zeta) = \lim_{t^P \to 0^+} \frac{1}{1 + e^{\frac{-\zeta}{t^P}}} = \begin{cases} 0 \text{ for } \zeta < 0 \\ \frac{1}{2} \text{ for } \zeta = 0 \\ 1 \text{ for } \zeta > 0 \end{cases} \tag{19}$$

In the DBN framework, the features are identified by utilizing a set of RBM layers and MLP is exploited for carrying out the classification process. The arithmetical representation is given by Eqs. (20) and (21), which indicates the energy of Boltzmann machine by arranging the neuron states that is usually binary bi and $L_{a,l}$ refers to the weights of related neurons θ_a points out the biases.

$$\mathrm{EN(bi)} = -\sum_{a<l} bi_a L_{a,l} - \sum_a \theta_a bi_a \tag{20}$$

$$\Delta\mathrm{EN(bi}_a) = \sum_l bi_a L_{a,l} + \theta_a \tag{21}$$

The novel definitions of energy for joint composition of hidden and visible neurons (x, y) are specified by Eqs. (22), (23), and (24), in which x_a and y_a specify the binary state of the visible unit, a and hidden unit l. k_a and C_l correspond to the biases, and $L_{a,l}$ refers to the weight among them.

$$\mathrm{EN}(x, y) = -\sum_{(a,l)} L_{a,l} x_a y_l - \sum_a k_a x_a - \sum_l C_l y_a \tag{22}$$

$$\Delta\mathrm{EN}(x_a, \bar{y}) = \sum_l L_{al} y_l + k_a \tag{23}$$

$$\Delta\mathrm{EN}(\vec{x}, y_a) = \sum_l L_{al} x_a + C_l \tag{24}$$

From the probability distribution of the input data, the weight constraints are encoded for training the RBM to utilize unsupervised learning. In addition, RBM is subjected to training for increasing the product of probabilities fixed to the learning models, which generate the maximal probability by weight allocation $\hat{L}_{\left(\hat{G}\right)}$ as per Eq. (25).

$$\hat{L}_{\left(\hat{G}\right)} = \max_{\hat{L}} \prod_{\vec{x} \in N} c(\vec{x}) \tag{25}$$

The DBN model exploits the energy function for the entire pairs (hidden and visible vectors) to assign the probability by RBM framework as specified in Eq. (26), which PR^F points out the partition function extracted from supplementing energy of the existing neuron states.

$$c\left(\vec{x}, \vec{hi}\right) = \frac{1}{\mathrm{PR}^F} e^{-\mathrm{EN}(\vec{x}, \vec{y})} \tag{26}$$

The RBM probability assigned to a visible vector derived from adding the energy of the entire available hidden neuron states is given by Eq. (27).

$$\mathrm{PR}^F = \sum_{\vec{x},\vec{y}} e^{-EN(\vec{x},\vec{y})} \tag{27}$$

The CD learning is exploited for initializing the visible neurons with input data rather than arbitrarily assigning the visible neurons that resulted in the formation of the model distribution.

DBN model exploits CD learning model, whose steps are given below.

Step 1: Select the x training samples.
Step 2: Assess the probability of hidden neurons c_y by recognizing the product of a visible vector x and $\hat{L}$ weight matrix as $c_y = \sigma\left(x.\hat{L}\right)$ depending on Eq. (28), where σ denotes the activation function.

$$c(\vec{y}_l \to 1|\vec{x}) = \sigma\left(C_l + \sum_a x_a L_{a,l}\right) \tag{28}$$

Step 3: Scrutinize y states from c_y probabilities.
Step 4: Compute x and c_y which is evaluated as a positive gradient $\phi^+ = x.c_y^{t^P}$.
Step 5: Analyze the x' visible states from y as shown in Eq. (29). In addition, it is necessary to assess y' from the renovation of x'.

$$c(\vec{x}_l \to 1|\vec{y}) = \sigma\left(k_a + \sum_a x_l L_{a,l}\right) \tag{29}$$

Step 6: Compute the exterior product of x' and y', by its gradient $\phi^- = x'.y'^{t^P}$.
Step 7: Describe the updated weight as given by Eq. (30), in which η denotes the learning rate.

$$\Delta\hat{L} = \eta\left(\phi^+ - \phi^-\right) \tag{30}$$

Step 8: Eq. (31) defines the weight update with novel values.

$$L'_{a,l} = \Delta L_{a,l} + L_{a,l} \tag{31}$$

Prior to the MLP learning process, consider $\left(K^{\hat{G}}, W^{\hat{G}}\right)$ training patterns, in which $\hat{G}$ indicates the counts of training patterns, and $1 \le \hat{G} \le \bar{P}$, $K^{\hat{G}}$, and $W^{\hat{G}}$ specify the input vector along with necessary output vectors, correspondingly. Every neuron error in l of output layer is demonstrated by Eq. (32). Consequently, Eq. (32) shows the squared error of $\hat{M}$ pattern as given in Eqs. (33) and (34), in which $\bar{P}$ symbolizes the count of training patterns.

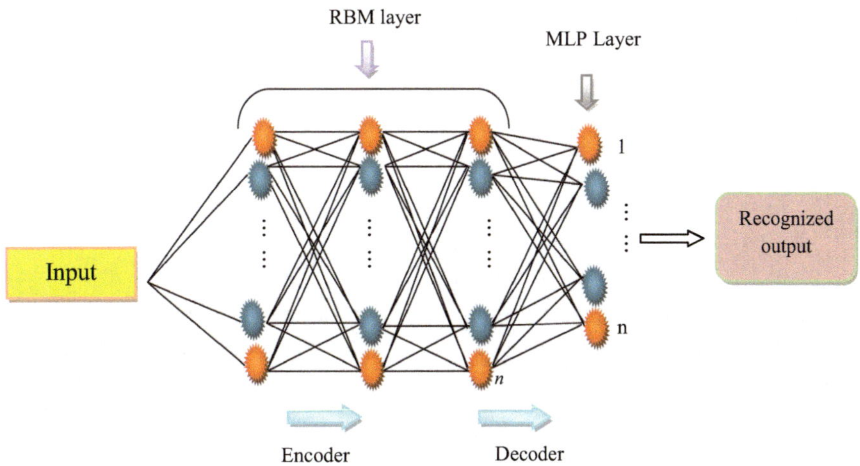

Fig. 5 Overall representation of the DBN model

$$e_l^{\hat{G}} = K^{\hat{G}} - W^{\hat{G}} \tag{32}$$

$$\mathrm{SE}_{\hat{G}}^{\mathrm{mean}} = \frac{1}{\tilde{o}_y}\sum_{l=1}^{\tilde{o}_y}\left(e_l^{\hat{G}}\right)^2 = \frac{1}{\tilde{o}_y}\sum_{l=1}^{\tilde{o}_y}\left(K^{\hat{G}} - W^{\hat{G}}\right)^2 \tag{33}$$

$$\mathrm{SE}_{\mathrm{avg}} = \frac{1}{\tilde{P}}\mathrm{SE}_{\hat{G}}^{\mathrm{mean}} \tag{34}$$

The two-phase training of DBN framework comprises of RBM and MLP, in which RBM exploits unsupervised learning, and MLP exploits the supervised learning. Figure 5 shows the model of DBN framework.

3.6 Solution Encoding

The DBN model includes three hidden layers with a set of hidden neurons. The count of hidden neurons in layer 1 is denoted by NH_1, layer 2 is as NH_2, and layer 3 is as NH_3 where both NH_1 and NH_3 are maintained constant. Rather than this, the count of hidden neurons in layer 2 NH_2 is selected optimally by concerning the objective function specified in Eq. (35). Figure 6 demonstrates the solution given to the D-MBO algorithm. Equation (37) depicts the objective function of this study, in

Fig. 6 Solution encoding

which e_y^k denotes the error among predicted and actual outcomes.

$$\text{Obj} = \text{Min}\left(e_y^k\right) \tag{35}$$

3.7 Conventional MBO

The migration operation of the monarch butterflies is used to solve multiple optimization issues with certain characteristics such as (i) the overall population of monarch butterflies is limited to Land 1 and Land 2. (ii) The migration operator produces every single child monarch butterfly in both the regions such as Land 1 and Land 2. (iii) In the MBO approach [11], the parent butterfly will pass away when a new one is created so as to maintain a constant population. (iv) The quality and performance of the monarch butterfly are assured even with the increased generations.

Migration Operator [27]: Let *MP* be the total number of the population, Maxgen specifies the maximum generation and c indicates the ratio of butterflies in Land a and Land b. For a better understanding, Land a and Land b are renamed as subpop 1 and subpop 2.

$$a_{i,y}^{m+1} = a_{s1,y}^{m} \tag{36}$$

Equation (42) defines the process of migration, where $a_{i,y}^{m+1}$ indicates the yth component of monarch butterfly i at generation $m + 1$ and $a_{s_1,y}^{m}$ specifies the yth component of a_{s_1} that is the new position of the monarch butterfly s_1. The variable m indicates the current generation. From subpop 1, monarch butterfly s_1 is randomly selected. The value s is defined as given in Eq. (37).

$$s = rn \times \text{time} \tag{37}$$

The variable *time* specifies migration time which is usually initiated based on 12 months of the year that can be given as 1.2. The variable *rn* represents an arbitrary value taken from uniform distribution. The component y for the newly produced monarch butterfly is initiated when $s > c$ as given in Eq. (38).

$$a_{i,y}^{m+1} = a_{s2,y}^{m} \tag{38}$$

In Eq. (40), $a_{s2,y}^{m}$ specifies the yth component of a_{s_2} which is the newly initiated location for the monarch butterfly s_2. From subpop 2, monarch butterfly s_2 is arbitrarily selected. On the basis of the above-mentioned analysis, by regulating the ratio c, the stability of the migration operator direction could be maintained in MBO model. The c value defines which subpopulation such as subpop 1 and subpop 2 is selected. The c value is 5/12 for the present calculation.

Butterfly Balancing Operator: In addition to the migration operator, a balancing operator is also used to update the location of the monarch butterflies. When an arbitrarily produced value *rn* is less than or equivalent to *c* for each component of the monarch butterfly *x*, the memory is then updated as shown in Eq. (41).

$$a_{x,y}^{m+1} = a_{\text{best},y}^{m} \tag{39}$$

where $a_{x,y}^{m+1}$ indicates the *y*th component of a_x for the generation $m+1$ that indicates the position of monarch butterfly *x*. Likewise, $a_{\text{best},y}^{m}$ specifies the *y*th component of a_{best} that indicates the best monarch butterfly in Lands *a* and *b*. In contrast, when *c* value is less than *rn*, the memory is updated as defined in Eq. (40), where $a_{s_3,y}^{m}$ indicates *y*th a component of a_{s_3} that is randomly selected in Land b. At this point, $s_3 \in \{1, 2, \ldots, MP_2\}$.

$$a_{x,y}^{m+1} = a_{s_3,y}^{m} \tag{40}$$

$$a_{x,y}^{m+1} = a_{x,y}^{m+1} + \lambda \times \left(fn_y - 0.5\right) \tag{41}$$

For the constraint, $rn > arbm$, the memory is then updated additionally as expressed in Eq. (41), in which *arbm* indicates a butterfly's balancing value. The variable *fn* represents the monarch butterfly *x* walk steps. It can be determined by using levy flight as defined in Eq. (42) and (43).

$$fn = \text{Levy}\left(a_x^m\right) \tag{42}$$

$$\lambda = H_{\text{dim}} \big/ m^2 \tag{43}$$

The weighting element λ is represented based on Eq. (43), where H_{dim} denotes the value that at a single move, a single monarch butterfly can move the maximum walk steps. When the value of λ is high, it represents the extended space of exploration, which maximize the impact of *fn* on $a_{x,y}^{m+1}$ which enhance the identification of more search space, whereas the value of λ is low, it specifies the short space of exploration, which minimize the impact of *fn* on $a_{x,y}^{m+1}$ which enhances the exploitation process. The pseudo-code of the conventional MBO model is given by Algorithm 1.

Algorithm 1: Conventional MBO approach

(continued)

(continued)

Start
Initialization: Initializing the generation counter $m = 1$. Initialize the population of MP single monarch butterflies arbitrarily; set the maximum generation hg, monarch butterfly number MP_1 in Land a
Monarch butterfly number MP_2 in Land b, max step H_{dim} butterfly balancing rate ardm, migration time *time*, and the migration ratio c
Fitness calculation: Estimation of every monarch butterfly based on its position
While $m < hg$ or if there is no optimal solution, do
Sort all the monarch butterfly singles corresponding to their fitness
Divide monarch butterfly singles into two subpopulations (Land a and Land b);
for $i = 1$ to MP_1 (for each monarch butterflies in subpop 1), do
Create a novel subpopulation 1 as per migration operator
end for i
for $j = 1$ to MP_2, do
Produce novel subpop 2 corresponding to Butterfly Balancing Operator
end for j
Combine the two currently generated subpopulations into one whole population
Estimate the population corresponding to the currently updated locations
$m = m + 1$
end while
Attain the optimal solution
End

3.8 Study on D-MBO Algorithm

Usually, MBO algorithm could deal with more benchmark optimization issues for discovering better operations. For making further enhancement on the performance of conventional MBO, this study adopts a new modified algorithm termed as D-MBO. Actually, the process of MBO is with two most important stages such as migration operator and butterfly adjusting operator, and in the conventional algorithm, the population is split into two arbitrary subpopulation. Instead of this, the introduced D-MBO algorithm divides the subpopulation depending on a distance function specified in Eq. (44), which is the distance among current solution and the best solution. Here, a_x and a_{best} refer to the current and best solutions, correspondingly.

$$\text{Dis} = a_{\text{best}} - a_x \tag{44}$$

The solution with the slightest distance is sorted, and the initial half set will be regarded as subpop 1, and subsequent half will be regarded as subpop 2. Therefore, the final best solution will be effectual that offers improved convergence. Despite this, this study makes a novel assessment on levy flight random walk. The step sizes (step_size) are produced by means of levy distribution for exploiting the search area and computed as per Eq. (45), in which ss refers to the weight and Max^{it} indicates the maximum iteration.

$$(\text{step\_size}) = ss \times \text{Max}^{\text{it}} \tag{45}$$

An innovative systematic valuation is carried out in this research work by varying the weighting factor ss to 0.5, 0.8, 1.0, 1.2, and 1.5, respectively. The pseudo-code of developed D-MBO is given in Algorithm 2, and its flow chart representation is given by Fig. 7.

Algorithm 2: D-MBO Algorithm

Begin
Initialization. Initializing the generation counter $m = 1$. Initialize the population of MP single monarch butterflies arbitrarily; set the maximum generation hg, monarch butterfly number MP_1 in Land a
Monarch butterfly number MP_2 in Land b, max step $H_{\dim}$ butterfly balancing rate ardm, migration time *time*, and the migration ratio c
Identify the distance among best solution and current solution using Eq. (44)
Sort the solution based on minimum distance
Select subpop 1 and subpop 2
Fitness Estimation. Estimation of every monarch butterfly based on its position
While the optimal solution is not found or $m < hg$ **do**
Sort all the monarch butterfly singles corresponding to their fitness
Divide monarch butterfly singles into two subpopulations (Land a and Land b)
For $i = 1\ to\ MP1$ (for each monarch butterflies in subpop 1) **do**
Generate new subpop 1 according to Migration operator
end for
for $x = 1\ to\ MP2$ (for each monarch butterflies in subpop 2) **do**
Produce novel subpop 2 corresponding to Butterfly Balancing Operator.
end for
Combine the two currently generated subpopulations into one whole population
Estimate the population corresponding to the currently updated locations
$m = m + 1$.
end while
Output the optimal solution.
End

4 Impact on Parameters

4.1 Simulation Setup

This study was implemented in MATLAB 2018a, and the performance of the study was noticed via the simulation outcomes. The database of DR diagnosis was gathered from the URL "[https://www5.cs.fau.de/research/data/fundus-images/] Access date: 2019-01-12)," which comprises of 15 images of healthy patients and 15 images of patients with DR. The study was carried out under diverse performance measures such as "accuracy, sensitivity, specificity, precision, FPR, FNR, NPV, FDR, F_1 Score,

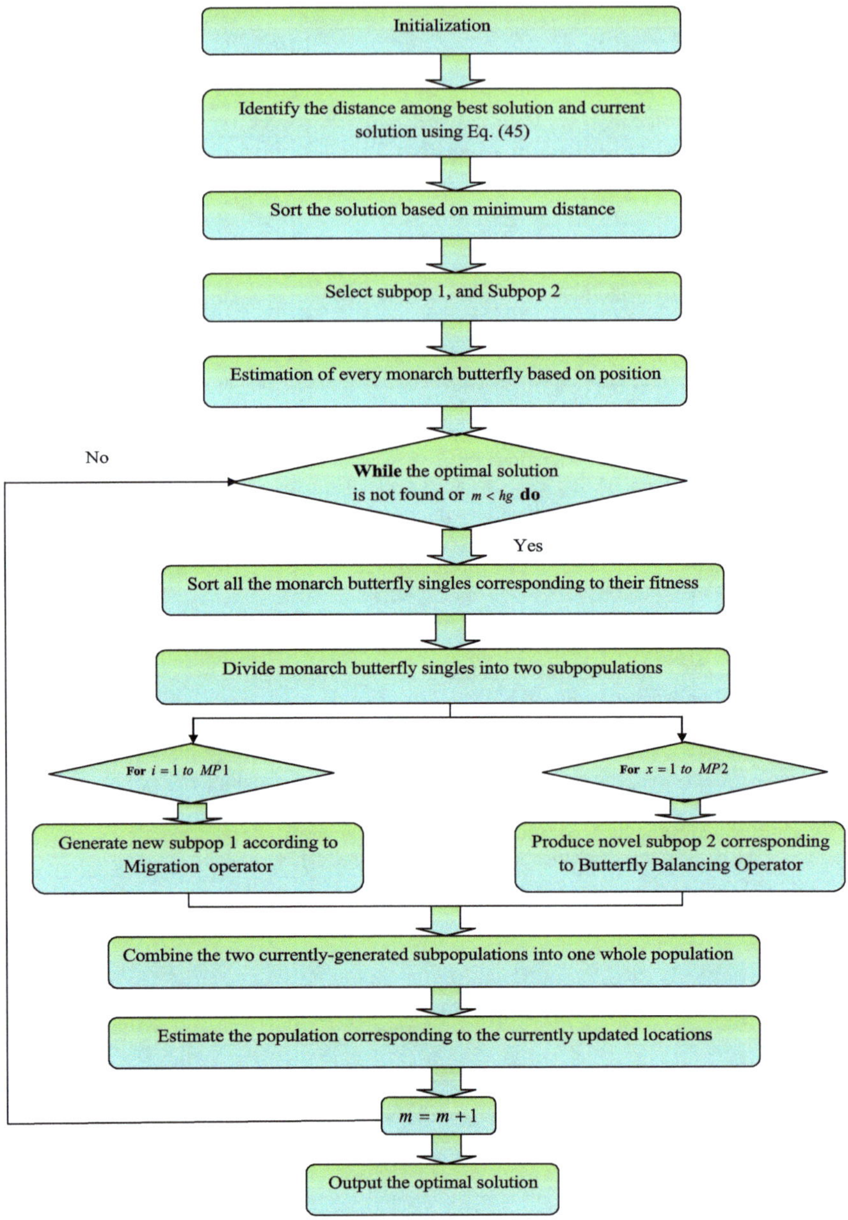

Fig. 7 Flow chart representation of the adopted study

Table 1 Overall performance analysis

Measure	$ss = 0.5$	$ss = 0.8$	$ss = 1.0$	$ss = 1.2$	$ss = 1.5$
Accuracy	0.66667	0.77778	0.55556	0.88889	0.66667
Sensitivity	0.5	0.66667	0.33333	0.83333	0.5
Specificity	1	1	1	1	1
Precision	1	1	1	1	1
FPR	0	0	0	0	0
FNR	0.5	0.33333	0.66667	0.16667	0.5
NPV	1	1	1	1	1
FDR	0	0	0	0	0
F1-score	0.66667	0.8	0.5	0.90909	0.66667
MCC	0.5	0.63246	0.37796	0.79057	0.5

and MCC." In this study, the analysis was held by varying the weighting factor and population size parameters of the adopted algorithm.

4.2 Overall Performance Analysis

Table 1 shows the overall performance attained by the adopted study by varying the weighting factor ss. From Table 1, higher accuracy rate was attained when $ss = 1.2$, which is 33.33, 59.99, and 14.28% better than the accuracy when $ss = 1.5$, 1.0, and 0.8, respectively. The sensitivity is very high under the variation $ss = 1.2$, which is 66.66, 60.00, and 24.99% better from the obtained sensitivity when $ss = 1.5$, 1.0, and 0.8, respectively. Likewise, the analysis is carried out for all the other measures under diverse variations. Thus, the outcomes have shown the betterment of this study.

4.3 Impact on Population Size

This division explains the algorithmic analysis of this study by changing the weighting factor ss of step size assessment under the levy flight distribution given in Eq. (45). The variation is under 0.5, 0.8, 1.0, 1.2, and 1.5, correspondingly. The study is carried out with respect to all the performance measures as specified in Fig. 8. It is noticed that the accuracy is high when $ss = 0.8$, which is 10.01, 23.41, and 55.13% improved from the accuracy rate when $ss = 1.2$, 1.0, and 1.5, correspondingly (for population size = 50). For population size = 60, this study attains higher accuracy rate when $ss = 0.8$, which is 13.22, 38.68, and 25.73% improved than the performance when ss is 1.2, 1.5, and 0.5, correspondingly.

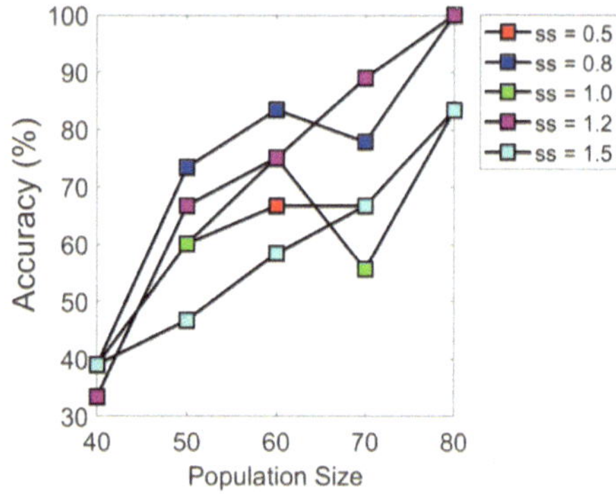

Fig. 8 Performance analysis by varying the population size in terms of accuracy

Likewise, the sensitivity shown in Fig. 9 is also examined for all the population sizes (40, 50, 60, 70, and 80). For population size = 60, the sensitivity is very high when $ss = 0.8$, and it is 23.56, 54.10, and 79.08% better from the performance when $ss = 1.2$, 0.5, and 1.5 in that order. For population size = 70, the study attains high sensitivity when $ss = 1.2$, which is 27.42, 59.01, and 57.36% better than the sensitivity when $ss = 0.8$, 1.5, and 1.0, respectively.

In addition, from Fig. 10, the specificity of the adopted study at $ss = 0.8$ is found to be higher when population size = 85, which is 35.29% and 18.1% better from the performance when $ss = 1.0$ and 1.5, respectively.

Moreover, from Fig. 11, the precision of the adopted study is very high when $ss = 0.8$, and it is 18.84%, 26.83%, 14.63%, and 39.02% better from the performance when $ss = 0.5$, 1.0, 1.2, and 1.5, respectively.

On examining the negative measures, the adopted study attains less FNR, FPR, and FDR under different variations of *ss*. From Fig. 12, the least FPR is attained by the presented study when $ss = 0.8$, which is 66.67 and 50% improved than the performance when *ss* is 1.0 and 1.5 correspondingly under population size 40.

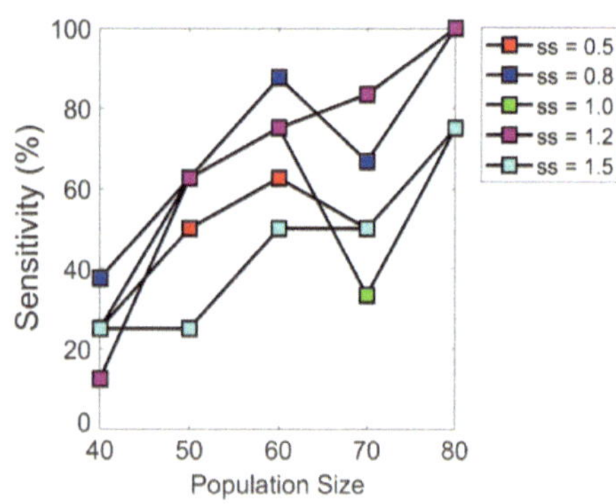

Fig. 9 Performance analysis by varying the population size in terms of sensitivity

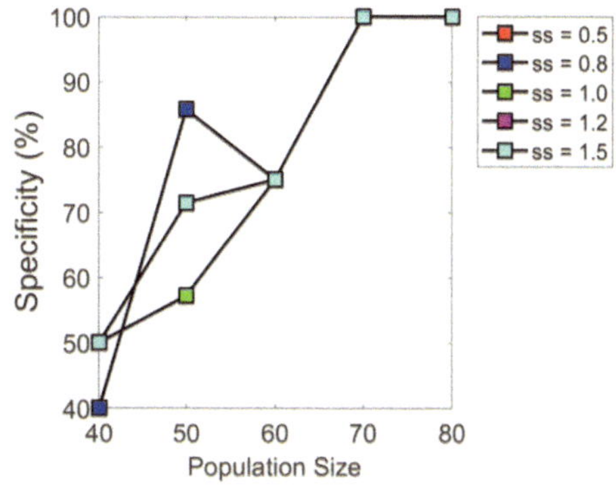

Fig. 10 Performance analysis by varying the population size in terms of specificity

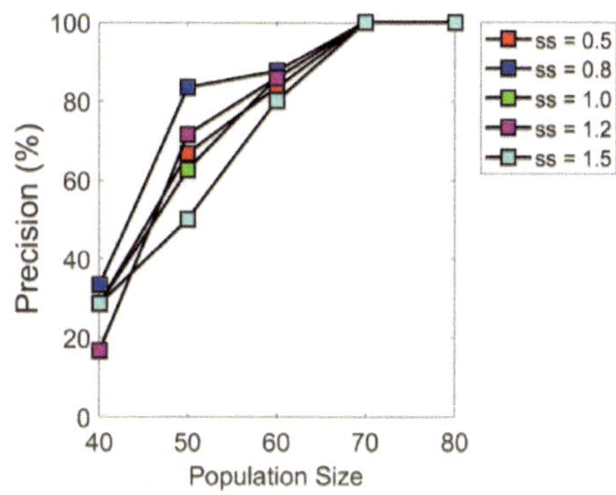

Fig. 11 Performance analysis by varying the population size in terms of precision

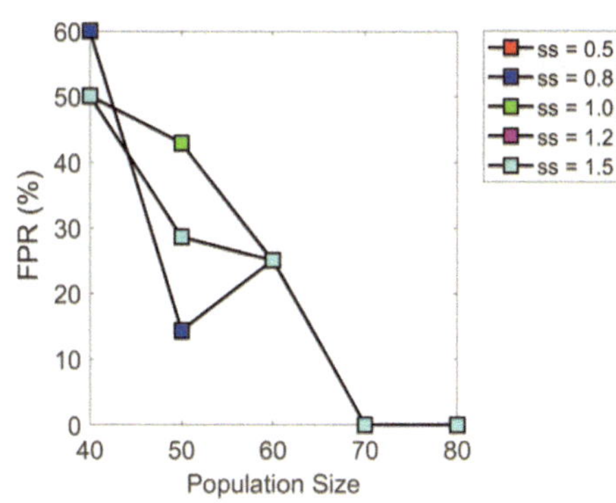

Fig. 12 Performance analysis by varying the population size in terms of FPR

In addition, from Fig. 13, the FNR measure attained by the study is lower when $ss = 0.8$ and it is 62.5, 66.67, and 66.67% better from the performance when $ss =$ 0.5, 1.5, and 1.2 in that order population size = 60.

In addition, from Fig. 14, the NPV of the adopted study at $ss = 0.8$ is found to be higher when population size = 85, which is 35.29% and 18.1% better from the performance when $ss = 1.0$ and 1.5, respectively.

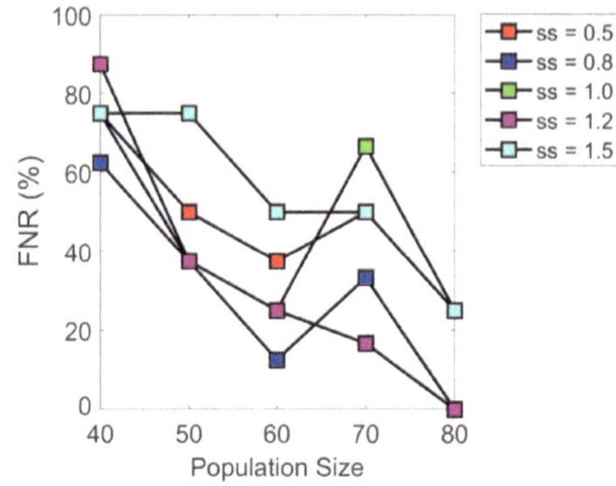

Fig. 13 Performance analysis by varying the population size in terms of FNR

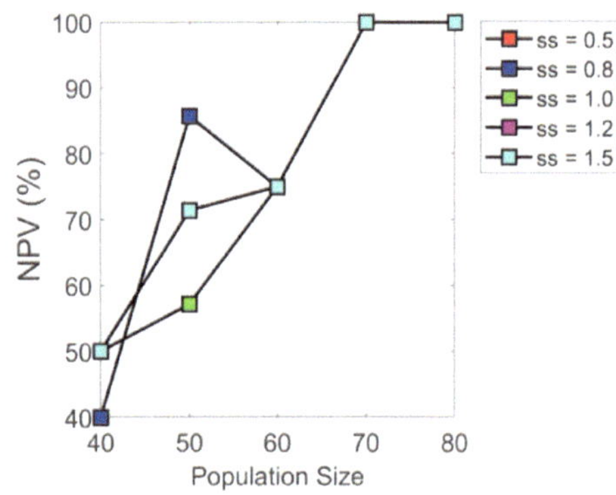

Fig. 14 Performance analysis by varying the population size in terms of NPV

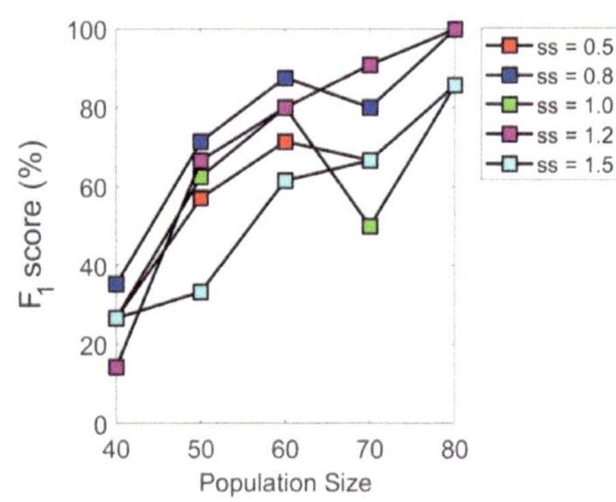

Fig. 15 Performance analysis by varying the population size in terms of F1-score

Likewise, the F1-score shown in Fig. 15 is also examined for all the population sizes (40, 50, 60, 70, and 80). For population size = 60, the F1-score is very high when $ss = 0.8$, and it is 23.56, 54.10, and 79.08% better from the performance when $ss = 1.2$, 0.5, and 1.5 in that order. For population size = 70, the study attains high F1-score when $ss = 1.2$, which is 27.42%, 59.01%, and 57.36% better than the NPV when $ss = 0.8$, 1.5, and 1.0, respectively.

Fig. 16 Performance analysis by varying the population size in terms of FDR

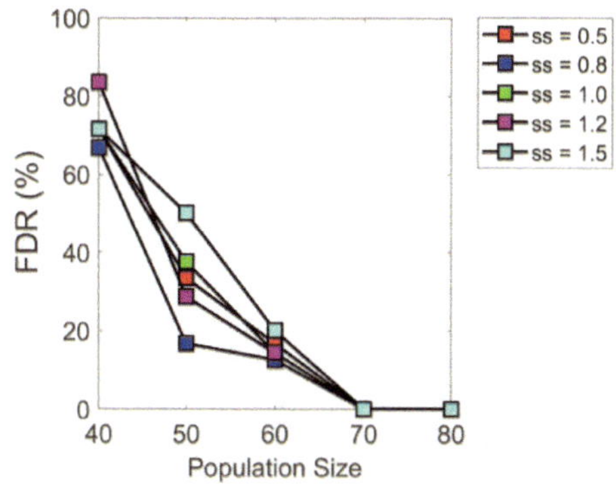

Fig. 17 Performance analysis by varying the population size in terms of MCC

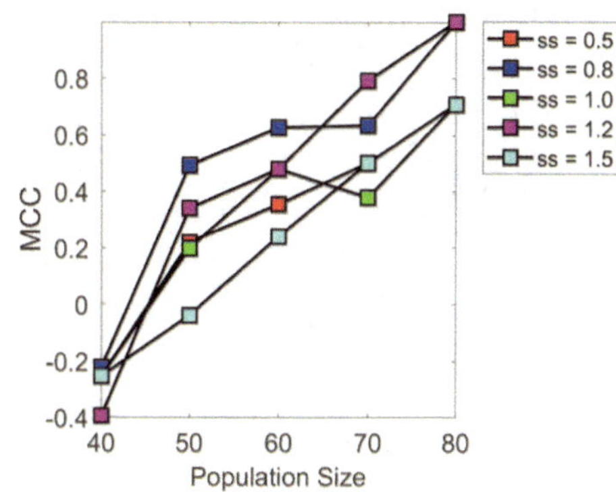

From Fig. 16, the FDR attained by the study when population size = 50 is lesser at $ss = 0.8$ and it is 64, 51.28, 57.89, and 45.71% better from the performance when $ss = 1.5$, 1.0, 1.2, and 0.5, correspondingly.

The MCC attained by the study is observed in Fig. 17, which is higher when $ss = 1.2$, and it is 43.75, 33.33, 52.5, and 43.75% better from the performance when $ss = 0.5$, 0.8, 1.0, and 1.5 correspondingly at 70th population size.

5 Conclusion

This study has explained the DR detection model that includes four phases, namely (i) pre-processing, (ii) blood vessel segmentation, (iii) feature extraction, and (iv) classification. The CLAHE and median filtering methods for pre-processing phase were described. For segmentation purpose, FCM thresholding was deployed, and subsequently, local and morphological transformation-oriented features were evaluated. Finally, DBN-oriented classification model was clearly studied on clarifying how the classification process works under the model. Moreover, the new optimization algorithm that exploited for enhancing the classification model was also examined. Finally, the study has extended to make the algorithmic evaluation. From the analysis, the accuracy was high when $ss = 0.8$, which was 10.01, 23.41, and 55.13% improved from the accuracy rate when $ss = 1.2$, 1.0, and 1.5, correspondingly. The FDR attained by the study when population size = 50 was lesser at $ss = 0.8$, and it was 64, 51.28, 57.89, and 45.71% better from the performance when $ss = 1.5$, 1.0, 1.2, and 0.5, correspondingly.

References

1. Ram K, Joshi GD, Sivaswamy J (2011) A successive clutter-rejection-based approach for early detection of diabetic retinopathy. IEEE Trans Bio Eng 58(3):664–673
2. Walter T, Klein J-C, Massin P, Erginay A (2002) A contribution of image processing to the diagnosis of diabetic retinopathy-detection of exudates in color fundus images of the human retina. IEEE Trans Med Imaging 21(10):1236–1243
3. Li Q, Wang Y, Zhang J, Xu G, Xue Y (2010) Quantitative analysis of protective effect of erythropoietin on diabetic retinal cells using molecular hyperspectral imaging technology. IEEE Trans Biomed Eng 57(7):1699–1706
4. Roychowdhury S, Koozekanani DD, Parhi KK (2014) DREAM: diabetic retinopathy analysis using machine learning. IEEE J Biomed Health Inf 18(5):1717–1728
5. Yen GG, Leong W (2008) A sorting system for hierarchical grading of diabetic fundus images: a preliminary study. IEEE Trans Inf Technol Biomed 12(1):118–130
6. Seoud L, Hurtut T, Chelbi J, Cheriet F, Langlois JMP (2016) Red Lesion detection using dynamic shape features for diabetic retinopathy screening. IEEE Trans Med Imaging 35(4):1116–1126
7. Narasimha-Iyer H et al (2006) Robust detection and classification of longitudinal changes in color retinal fundus images for monitoring diabetic retinopathy. IEEE Trans Biomed Eng 53(6):1084–1098
8. Zhang B, Kumar BV, Zhang D (2014) Detecting diabetes mellitus and nonproliferative diabetic retinopathy using tongue color, texture, and geometry features. IEEE Trans Biomed Eng 61(2):491–501
9. Pires R, Avila S, Jelinek HF, Wainer J, Valle E, Rocha A (2017) Beyond lesion-based diabetic retinopathy: a direct approach for referral. IEEE J Biomed Health Inform 21(1):193–200
10. Kandemir Melih, Hamprecht Fred A (2015) Computer-aided diagnosis from weak supervision: a benchmarking study. Comput Med Imaging Graph 42:44–50
11. Desai Jay, Taylor Gretchen, Vazquez-Benitez Gabriela, Vine Sara, O'Connor Patrick J (2017) Financial incentives for diabetes prevention in a medicaid population: study design and baseline characteristics. Contemp Clin Trials 53:1–10
12. Bonanno Lilla, Marino Silvia, Bramanti Placido, Sottile Fabrizio (2015) Validation of a computer-aided diagnosis system for the automatic identification of carotid atherosclerosis. Ultrasound Med Biol 41(2):509–516
13. Acharya UR, Vidya KS, Ghista DN, Lim WJE, Sankaranarayanan M (2015) Computer-aided diagnosis of diabetic subjects by heart rate variability signals using discrete wavelet transform method. Knowl Based Sys 81:56–64
14. Fujita Hiroshi, Uchiyama Yoshikazu, Nakagawa Toshiaki, Fukuoka Daisuke, Zhou Xiangrong (2008) Computer-aided diagnosis: the emerging of three CAD systems induced by Japanese health care needs. Com Methods Prog Biomed 92(3):238–248
15. Berridge Emma-Jane, Roudsari Abdul, Taylor Sheila, Carey Steve (2000) Computer-aided learning for the education of patients and family practice professionals in the personal care of diabetes. Com Methods Prog Biomed 62(3):191–204

16. Amin J, Sharif M, Yasmin M, Ali H, Fernandes SL (2017) A method for the detection and classification of diabetic retinopathy using structural predictors of bright lesions. J Comput Sci 19:153–164
17. Franklin SW, Rajan SE (2014) Diagnosis of diabetic retinopathy by employing image processing technique to detect exudates in retinal images. IET Image Process 8(10):601–609
18. Zhang L, Li Q, You J, Zhang D (2009) A modified matched filter with double-sided thresholding for screening proliferative diabetic retinopathy. IEEE Trans Inf Technol Biomed 13(4):528–534
19. Agurto C et al (2010) Multiscale AM-FM methods for diabetic retinopathy lesion detect. IEEE Trans Med Imaging 29(2):502–512
20. Akram MU, Khalid S, Khan SA (2013) Identification and classification of microaneurysms for early detection of diabetic retinopathy. Pattern Recogn 46(1):107–116
21. Basha S Algorithmic analysis of distance based monarch butterfly oriented deep belief network for diabetic retinopathy. In Communication
22. Chang Y, Jung C, Ke P, Song H, Hwang J (2018) Automatic contrast-limited adaptive histogram equalization with dual gamma correction. IEEE Access 6:11782–11792
23. Zhu Youlian, Huang Cheng (2012) An improved median filtering algorithm for image noise reduction. Phys Procedia 25:609–616
24. Masood A, Jumaily AAA, Hoshyar AN, Masood O (2013) Automated segmentation of skin lesions: Modified fuzzy C mean thresholding based level set method. INMIC, Lahore, pp 201–206
25. Zhu C, Zou B, Zhao R, Cui J, Duan X, Chen Z, Liang Y (2017) Retinal vessel segmentation in colour fundus images using extreme learning machine. Comput Med Imagin Graph 55:68–77
26. Wang HZ, Wang GB, Li GQ, Peng JC, Liu YT (2016) Deep belief network based deterministic and probabilistic wind speed forecasting approach. Appl Energy 182:80–93
27. Wang GG, Deb S, Cui Z (2015) Monarch butterfly optimization. Neural Comput Appl 1–20
28. Niemeijer M, Abramoff MD, van Ginneken B (2009) Information fusion for diabetic retinopathy CAD in digital color fundus photographs. IEEE Trans Med Imaging 28(5):775–785

Chapter 10
A Modified Blind Deconvolution Algorithm for Deblurring of Colored Images

Anuj Kumar Gupta, Manvinder Sharma, Sohni Singh and Pankaj Palta

1 Introduction

Use In still cameras and video surveillance cameras, to reduce the shaking effects the optically stabilized lenses are used. These lenses are expansive. To keep optical assembly of camera steady, these system uses set of inertial sensors and gyroscope. But these are effective only for small camera shake at relatively small exposures of about 1/15th second. When there is no information of type of distortion, noise or blur in the image, the blind deconvolution works efficiently [1].

The estimation of the blurred kernels in the blurred image is a very difficult problem [2–5]. When the image has a camera rotation or a dynamic scene in it, then the estimation of the noise becomes very difficult because it becomes spatially invariant. Eliminating the noise from the image is the second challenge for the recovery of a noise-free image. Attenuation of the information of high frequency from the scene and averaging the neighboring pixels is done by noise. The algorithms or deblurring systems are needed to address the problem to recover the image free of noise [6–8].

There a variety of sources through which the blurring of images is aroused which include defocus of lens, temporal and spatial integration of sensor, scattering in the atmosphere, and optical aberration [9]. The visual systems of humans perceive it very quickly though its mechanism is partially understood. Hence, the estimation of the blur images becomes quite difficult [10, 11]. Blur is basically created because of the movement of the camera, and inaccurately focusing it or due to the creation of

A. K. Gupta (✉) · M. Sharma · S. Singh · P. Palta
CGC College of Engineering, Landran, Mohali, India
e-mail: anuj.coecse@cgc.edu.in

M. Sharma
e-mail: manvinder.coeece@cgc.edu.in

P. Palta
e-mail: pankaj.coecse@cgc.edu.in

S. Jain et al. (eds.), *Advances in Computational Intelligence Techniques*,
Algorithms for Intelligent Systems,
https://doi.org/10.1007/978-981-15-2620-6_10

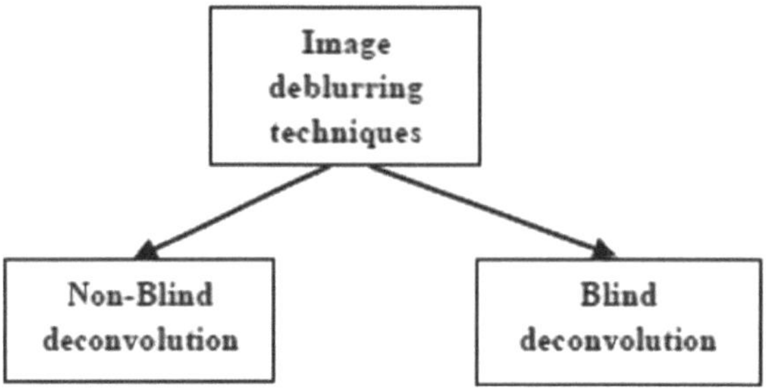

Fig. 1 Types of image deblurring techniques

shallow field depth caused by the use of an aperture resulting in an image area which is nonsharp making it blur in nature [12] (Fig. 1).

Whenever we have no information regarding the blurring and noise that produces distortion, we can effectively implement the blind deconvolution algorithm [13–16]. But for achieving high efficiency and quick estimation of the motion blur and high speed, no doubt these methods and algorithms restore images of very high resolution which makes them useful for photographers as a post-processing tool. The neural networks are being explored in this work in comparison to the traditional methods where there are multiple refinements to produce the restored image. The insight into the neural architecture design for deconvolution process is provided. The neural network used in the core of the algorithm differs from the architectures used previously in many significant ways. The neural networks use complex Fourier coefficients of deconvolution filter which is applied at the input. The input patch is decoded using frequency decomposition at multiple resolutions. The number of weights in the neural network training is significantly reduced which allows the successful training of the neural network for deblurring the images using blur kernels at very large patches. Figure 2 shows the basic block diagram of neural blind deconvolution.

An input patch is given which is blurred and blur kernel is unknown. The Fourier coefficients are predicted by the neural network which is further applied for the restored output. The input image contains a large number of overlapping patches which are having unknown blur kernels. These input patches are applied to the neural network independently and the outputs are composed to form the sharp image estimates. A blur kernel is then inferred which is related to the initial estimate and this kernel is then used for non-blind deconvolution [17]. This approach is quite accurate and robust.

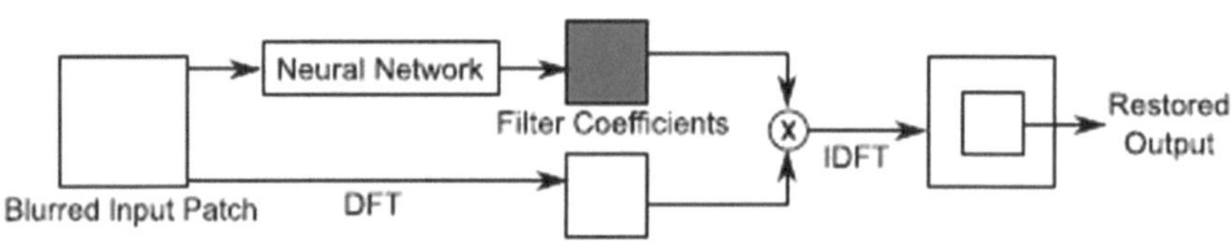

Fig. 2 Neural blind deconvolution

2 Literature Review

The restoration of the blurred image is basically obtaining a clear image or the original one which has been corrupted by noise or any other disturbances. These disturbances come in many forms such as motion blur, defocusing of the camera or noise which may blur the original image. Regaining back the information of the image which is lost during the blurring process is image restoration. There are many image enhancement techniques which defocus the image and emphasize the particular characteristics of the image but image enhancement is quite different from image restoration. Image enhancement could not be used in quite a few applications because it could not produce the necessary realistic data scientifically required. The noise could be removed through this method but by compromising the resolution of the image which was not acceptable for many applications. But in image restoration, the deblur image reduces the noise and recovers the resolution of the image as well. The use of neural networks for restoring the text images in blind deconvolution is used in [18]. Since there are sparse contours present in the highly structured images, a feed-forward architecture was used for successfully restoring the blur images in [19] a neural network was trained successfully to identify the blur where there were restrictions on the motion of the blur.

3 Neural Deconvolution and Restoration

Let $b[n]$ be the blurred image due to camera motion and $a[n]$ be the corresponding sharp image that is estimated. The convolution indicates the degradation obtained due to blur with a blur kernel k which is unknown.

$$b[n] = (a * k)[n] + \varepsilon[n], k[n] \geq 0 \sum_{n} k[n] = 1 \tag{1}$$

Here, * denotes convolution and $\varepsilon[n]$ is the Gaussian noise.

In the algorithm, the neural network is the central component and the restoration on individual patches $b[n]$ is carried out. The design of the network is done in a way such that it is able to recover the values of sharp intensity of patches p.

$$a_p = \{a[n] : \mathrm{n} \in \mathrm{p}\} \tag{2}$$

A larger patch b_{p+} from the observed image is given as:

$$b_{p+} = \{a[n] : n \in p^+\}, p^+ \supset p \tag{3}$$

The discrete Fourier transform (DFT) coefficients of the deconvolution filter are the output of the neural network show in Fig. 2.

$$A'_{p+}[z] = G_{p+}[z] \cdot B_{p+}[z] \tag{4}$$

By taking the inverse discrete Fourier transform (IDFT) of $A'_{p+}[z]$, we get $a'_p[n]$ of the sharp image. The main objective of the neural network training is that $G_{p+}[z]$ should be optimal with respect to $a_p[n]$. In other words, the output coefficients should be optimal with respect to the quality of the image with sharp intensities. The mean square error (MSE) is defined as the difference between the predicted value $a'_p[n]$ and the true sharp image intensity $a_p[n]$. The loss function of the neural network is defined by MSE.

$$M\left(a'_p, a_p\right) = \frac{1}{|p|} \sum_{n \in p} \left(x' p[n] - xp[n]\right)^2 \tag{5}$$

The synthetic database is used to train the neural network. The patches of the sharp image are extracted from the image and then they are blurred with the kernels which are synthetically motion generated and the Gaussian noise is further added. The noise is set at 1% standard deviation so that the noise level present at the benchmark is matched [20]. Random points are sampled for generating the kernels with synthetic motion and then a spline is fit into these points. The kernel values present at each pixel is set to a value which has Gaussian distribution of mean one and half standard deviation.

4 Neural Architecture

In order to handle large blur kernels, the network must be able to process large input patches. The training provided to the neural network should be quite feasible so that large number of weights could be learned [17]. The input patch is parameterized and the initial layers of the network are connected to each other. The decomposition strategy at multiple resolutions is used in Fig. 3. Here, sampling between the higher spatial frequencies takes place with lower resolution. The DFT is computed at three different levels. These levels correspond to the different size patches.

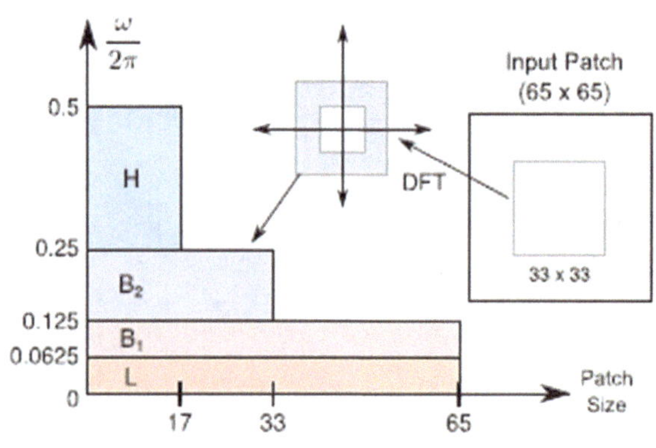

Fig. 3 Multi-resolution decomposition

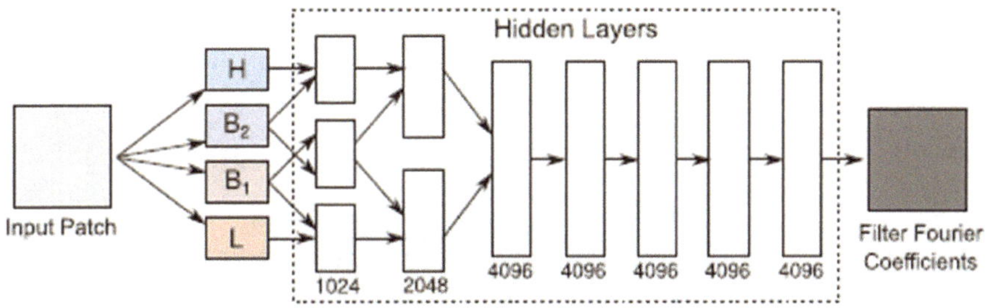

Fig. 4 Feed-forward network architecture

This is done so that the number of weights used in the network is minimized. The input patch is encoded into four bands which contain frequency components of low pass, i.e., L, two band-pass, i.e., B_1 and B_2, and H as high pass frequency component. The sampling of higher frequencies takes place at the coarser resolution. A small patch is centered at the input and all the sampled higher frequencies are calculated here. After the decomposition, we get complex coefficients which are independent in nature from the DFT that were grouped into bands. A feed-forward architecture of network is shown in Fig. 4.

The hidden layers in the network architecture predict the complex coefficients obtained from DFT. The blurred input patch is encoded to obtain the coefficients. The input coefficients are connected to the first layer of the network architecture from the frequency bands that are adjacent. The weights are not shared by these groups. Similar type of strategy is adopted for the next layer. The adjacent groups present in the first layers are connected by the units. The connectivity is restricted based upon the frequency localization. In this way, the number of weights in the network is reduced while still practicing good prediction. In many iterative algorithms, the task is divided sequentially at individual scales [21, 22]. There is ReLU activation in all the hidden layers [23]. Then the blurred kernels of different sizes having random synthetic motion are used to train the neural network.

5 Design Flow

The flowchart shows the steps for deblurring the images. Firstly, the blurred image is read, then restoring the blurred image using the point spread function for different sizes. Then analysis of restored PSF is analyzed and improving of restoration is done. The references are read from the colored image for detection in the first step. Then in the next step, the point spread function (PSF) of the image is done. Then the colored image is converted into the undersized PSF which is 4 times pixels shorter. And then again this colored image is converted into the oversized PSF which is 4 times pixels higher than the initial one. And then the colored image is converted into the initial PSF which is of the same pixel size. The PSF images which are stored are analyzed.

The edge detection using Red, Green, and Blue of the colored image is done using Sobel filter. The last step includes the deblurring of the colored image using blind deconvolution for each Red, Green, and Blue. These are the steps for deblurring the image using blind deconvolution method.

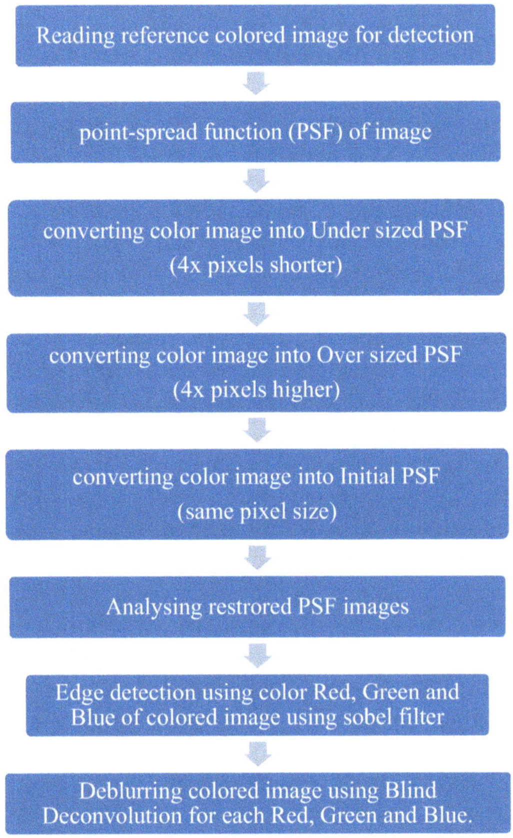

6 Results

The modified algorithm has novel approach to deblur colored images when any type of distortion is not known. The algorithm uses blind deconvolution after analyzing the image using PSF and damped and accelerated Richardson–lucy algorithm is used in all the iterations. The approach is capable to deblur colored images as well. To show the effectiveness of algorithm, the blurred color image is taken as example which is shown in Fig. 5. The point spread function is used on taken image, and undersized psf function, i.e., 4 times less pixel is applied on input image. The result is shown in Fig. 6. The oversized psf fuction is applied with 4 times pixel of image size. The

Fig. 5 Blurred image

Fig. 6 Deblurring with undersized (−4× pixel) PSF

result is shown in Fig. 7. The edges are detected of image and shown in Fig. 8. Color Sobel filter (Red, Green, and Blue) are applied on the edges detected in the image to reproduce the results. Analysis of restored PSF is done and using additional constrains, the PSF is restored. Figures 11, 12, and 13 shows the reproduced deblurred color image. The image id is restored in Figs. 11, 12, and 13. With the modified algorithm, the blurred image is reproduced and it can be observed that the image restoration with blue Sobel filter provides better deblurred image. The algorithm works well for colored images as well as grayscale images (Figs. 9 and 10).

Fig. 7 Deblurring with oversized (4× pixel) PSF

Fig. 8 Edge detection of colored image

Fig. 9 Reconstruction using Sobel filter for Red, Green, and Blue

Fig. 10 Application of Sobel filter on Red, Green, and Blue channels on detected edges

Fig. 11 Reproduced deblurred image for green channel

Fig. 12 Reproduced deblurred image for red channel

Fig. 13 Reproduced deblurred image for blue channel

7 Conclusion

In this paper, a modified form of restoring blurred images is shown. The images are restored by training a neural network that follows the deconvolution of the blur kernels to obtain the sharp images. The restored image has the quality much better than the algorithm used previously. There are different techniques available to restore the blurred image. Many algorithms are there which uses filter to restore the blur image. Through the approach of neural network, the deconvolution is done with the Fourier coefficients obtained through DFT at different stages corresponding to different patches of the input image. The Gaussian noise is also added. The neural network is trained to design the algorithm that produces image very close to the true image. The mean square error of the image is also calculated to infer the losses.

References

1. Kundur D, Hatzinakos D (1996) Blind image deconvolution. IEEE Signal Process magazine, 13(3):43–64
2. Lui G, Chang S, Ma Y (2014) Blind image deblurring using spectral properties of convolution operators. IEEE Trans Image Process 23(12):5047–5056
3. Queiroz F, Ren T, Shapira L, Banner R (2013) Image deblurring using maps of highlights. In: International conference on acoustics, speech and signal processing, pp 1608–1611
4. Brostow GJ, Essa I (2001) Image-based motion blur for stop motion animation. In: Fiume E (ed) SIGGRAPH 2001, computer graphics proceedings, pp 561–566
5. Bertero M, Boccacci P (1998) Introduction to inverse problems in imaging. Institute of Physics Publishing, Bristol and Philadelphia
6. Ben-Ezra M, Nayar SK (2004) Motion-based motion deblurring. IEEE Trans Pattern Anal Mach Intell 26(6):689–698

7. Jiang X, Cheng DC et al (2005) Motion deblurring. Department of Mathematics and Computer Science, University of Muenster
8. Kubota A, Aizawa K (2002) Arbitrary view and focus image generation: rendering object-based shifting and focussing effect by linear filtering. In: International conference on ICIP02, pp 489–492
9. Kang S, Min J, Paik J (2001) Segmentation-based spatially adaptive motion blur removal and its application to surveillance systems. International conference on ICIP01, pp 245–248
10. Cash BR, Liary DP (2015) Gide: graphical image deblurring exploration. Copublished by the IEEE CS and the AIP, University of Maryland
11. Banham MR, Katsaggelos AK (1997) Digital image restoration. IEEE Signal Process Magazine 14(2):24–41
12. Gonzalez RC, Woods RE (2002) Digital image processing. Pearson Education, London
13. Biretta J (1994) WFPC and WFPCC 2 instrumental characteristics, in the restoration of HST images and spectra-2. Space Telescope Science Institute, Baltimore, MD, pp 224–235
14. Cannon M (1976) Blind deconvolution of spatially invariant image blurs with phase. IEEE Trans Acoust Speech Signal Process 24(1):58–63
15. Savakis AE, Trussell HJ (1993) Blur identification by residual spectral matching. IEEE Trans Image Process 2(2):141–151
16. Ayers GR, Dainty JC (1988) Iterative blind deconvolution method and its application. Optics Lett 13(7):547–549
17. Xu L, Ren JS, Liu C, Jia J (2014) Deep convolutional neural network for image deconvolution. In: NIPS
18. Hradiš M, Kotera J, Zemcık P, Šroubek F (2015) Convolutional neural networks for direct text deblurring. In: Proceedings of BMVC
19. Sun J, Cao W, Xu Z, Ponce J (2015) Learning a convolutional neural network fornon-uniform motion blur removal. In: Proceedings of CVPR
20. Levin A, Weiss Y, Durand F, Freeman WT (2009) Understanding and evaluating blind deconvolution algorithms. In: Proceedings of CVPR
21. Michaeli T, Irani M (2014) Blind deblurring using internal patch recurrence. In: Fleet D, Pajdla T, Schiele B, Tuytelaars T (eds) ECCV 2014, vol 8691. LNCS. Springer, Heidelberg, pp 783–798
22. Sun L, Cho S, Wang J, Hays J (2013) Edge-based blur kernel estimation using patch priors. In: Proceedings of ICCP
23. Nair V, Hinton GE (2010) Rectified linear units improve restricted Boltzmann machines. In: Proceedings of ICML

Chapter 11
Assessment of Health Care Techniques in IoT

Chander Diwaker, Ajay Jangra and Ankita Rani

1 Introduction

While the systems of related devices are creating and extending, the Internet of Things (IoT) is wandering into each part of life. IoT gives a fit and organized way to deal with improves the health and success of mankind. The results of IoT advancements entering to restorative and healthcare parts are molding another sorted out correspondence course among parental figures and patients [1]. In such systems, a course of action of related wearable or implantable sensors consistently read patient's fundamental signs enabling parental figures to get to the data through the Internet [2].

IoT-enabled health observing systems often work such that remote body region organizes (WBAN), which are a lot of restorative sensors associated with the patient's body, records physiological parameters and critical signs and sends them to a cloud server for furthermore dealing with and limit [3]. Considering every sensor in WBAN as an IP-based related node and IoT-enabled healthcare system offers an open way to serve patients requiring steady observing outside emergency clinic conditions.

It has been represented that the amount of essentially wiped outpatients is creating and various patients leaving recuperating focus are still at the risk of rot at home [4]. A couple of those patients may encounter health disintegrating when there are some unpredictable changes giving up in their basic indications [5].

Remembering the ultimate objective to envision health debilitating, a technique called early cautioning score (EWS) has been proposed [6]. In this system, restorative orderlies record the patients' key signs in a perception diagram at a certain time between times and consign a score to the estimation of each sign in its range. The general patient score, which is the total of each individual score, is then used to choose whether the patient is self-destructing or not. The way toward account and figuring

C. Diwaker (✉) · A. Jangra · A. Rani
Department of Computer Science and Engineering, U.I.E.T., Kurukshetra University, Kurukshetra, Haryana, India

S. Jain et al. (eds.), *Advances in Computational Intelligence Techniques*, Algorithms for Intelligent Systems,
https://doi.org/10.1007/978-981-15-2620-6_11

EWS is commonly still paper-based and manual in recuperating offices. Paper-based and inexact information collection drives every now and again to the wrong estimation of caution score making health authorities to misdiagnose the circumstance. Furthermore, a manual paper-based methodology is moderate and requires work-time resources to outline the parental figures. As of late, early cautioning systems have begun to advance toward mechanized electronic stages. IoT-enabled wearable sensor systems can help the methodology of EWS computerization to be connected to in-home use cases [7].

Early warning system is a methodology for early recognition of health deterioration to constrain the impact of unexpected serious changes in health. Such a system uses a procedure called early warning score (EWS) to figure unmistakable scores from patient's observation layout in perspective on a redundant physiological estimation of vital signs to decide a composite score which is used to recognize if a patient is at a danger of disintegrating [8]. Gathers in this field have alluded to that patients often have clinical rot up to 24 h before a real clinical case requiring a full intervention. The most straightforward sort of score can be determined by utilizing physiological parameters: level of cognizance, beat rate, systolic circulatory strain, breath rate, body temperature, and blood oxygen submersion. Every parameter has the biggest score of 3 and a base score of 0 from which the last score can be discovered [9]. Lower scores lead to change in recognition repeat and higher scores results are giving a progressively raised measure of helpful keep an eye on the patient, for example, assessment by health ace or trading to the emergency unit.

2 System Architecture

The system design includes three principles as shown in Fig. 1:

(i) *Sensor Network*: In the sensor organize, remedial parameters are recorded by wearable sensors. Sensors are arranged into three social affairs in light of their

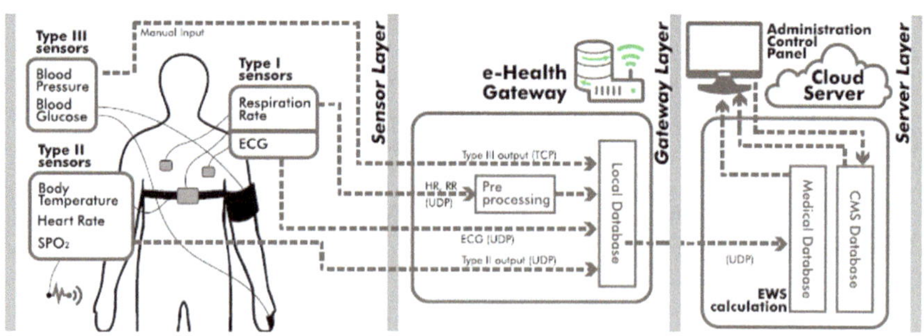

Fig. 1 IoT healthcare system [8]

data change rate and their ability. The essential social affair of sensors incorporates high date rate sensors used for spouting like steady parameters (e.g., ECG banner) [8]. The subsequent social affair included sensors that read and record data at a lower data rate, for example, body temperature and condition sensors. The last total fuses sensors which are not totally robotized and are sporadically used by patients or home caregivers. The regards perused by these sensors (e.g., circulatory strain) are added to the system physically [10].

(ii) *E-Health Gateway*: The entrance gets data from a few sensors, performs show change, and gives a few organizations, for instance, data weight and limit in offline mode. At the door layer, there can be two various sorts of methodologies depending upon the openness of the Internet association. In the event that the e-Health entryway has the Internet get to, it will send the progressing non-pre-taken care of signs (e.g., ECG signal) to the cloud server [8]. In this methodology, other data from the principle sensor create can be pre-dealt with and the numerical outcomes close by other data from the second kind sensors will be sent to the cloud server. In case the detachment of the Internet at the section, the e-Health gateway will store the compacted unrefined. At the point when the door is reconnected to the Internet, it will send first figured regards and after that packed unrefined data groups to the cloud server [11].

(iii) *Back-End System*: The back-end structure has two distinct areas:

(1) A cloud-based back-end establishment including information limit, data assessment, fundamental initiative, etc.,

(2) The User Interface (UI) which goes about as a dashboard for helpful care providers together with performing customer control and data recognition. The cloud server gets particular sorts of sensor information using looking at show and stores it in the database [12]. At this stage, early cautioning score is processable when the server has gotten every single bit of parameters required for estimation of EWS, and in perspective on the last score, authentic alerts will fire in case of emergency. A cloud server is in a like manner accountable for giving the organization control board to health pros with nonstop health information visualizes and UI for patients and in-home guardians [13].

3 Related Work

Ukiset al. [14] built an architecture of cloud-relied advanced medical image visualization solutions. Another idea was the early cautioning score (EWS), which was an approach to manage recognize the debilitating of a patient.

Rao et al. [15] proposed a modernized EWS prosperity watching structure to splendidly screen fundamental signs and envision prosperity disintegrating for in-home patients using Internet of Things (IoT) progressions. IoT enables our response to forgive a continuous throughout the day, consistently advantage for prosperity specialists to remotely screen in-home patients by methods for the Internet and get notifications if there ought to emerge an event of an emergency. They displayed a proof-of-thought EWS system where consistent scrutinizing, trading, recording, and treatment of key signs have been executed.

In light of the essentialness of watching the helpful state of patients who were encountering extraordinary disorders, especially cardiovascular contaminations, a steady remote patient checking was major. The checking structure should have the ability to give steady systematic organizations, desire, and alerts if there ought to be an event of an emergency for customers doing any development, wherever and at whatever point.

Ray et al. [16] presented an IoT-based approach to manage offer sharp remedial alerted in altered steady checking. The proposed methodology considers neighborhood preparing perspective enabled by AI figuring and automates the organization of structure sections in enlisting territory. The proposed structure was surveyed by methods for a logical examination concerning determined patient seeing to early recognize calm debilitating by methods for arrhythmia in ECG banner. The proposed methodology gave upgrades for the clients just as for the system. It empowered the customer and related parental figure to get close by notifications if there ought to be an event of present or not all that inaccessible future foreseen emergency condition.

Anzanpour et al. [17] presented the joint effort of IoT and cloud that offers profitable healthcare watching structures in which medicinal data can be moved securely with the consent of the patient. A system was worked among every one of the substances taking an interest in healthcare, which upgrades correspondence among the substances consequently passing on better care and organizations. Dissent hyper associating was another age advancement that targets loosening up the Internet to articles and areas around the globe. The huge information produced from different sources abides in the cloud, which requires more prominent getting ready vitality to recuperate data in a sheltered and solid way. An unavoidable observation system contains sensors, actuators, and cameras. In view of its bunch favorable circumstances, work topology was used as a piece of this reconnaissance system.

Azimi et al. [18] presented the continuous information was taken care of in the cloud server. Any change recognized in the information naturally makes all information be refreshed. This permits the concerned components related to offering healthcare to give the essential guidance and headings to the patients subsequent to inspecting the constant data. Data investigation was of essential significance in the productive acknowledgment of IoT devices and organizations. The atypical regions were the sections in test health parameter information that was of functional energy to the restorative professionals. This territory of research was of significant significance and was of a plenteous bit of leeway to mankind.

The healthcare investigation using a mind-boggling peculiarity disclosure engine would encourage early and induce acknowledgment of predominant sicknesses. The basics of IoT organizations include information protection, sensor information weight, and security. A persuading component was information quality, which was basic for continuous remote health watching. The primary test of inconsistency distinguishing proof was to confine the risk of the malady going undetected which was practiced through IoT [19–22].

The IoT empowered healthcare systems to have a high ground over the regular watching structures as the older were in consistent need of care. The IoT empowered checking structure uses a focal unit for the essential initiative. This focal fundamental authority unit can perceive perilous and basic circumstances depending upon the information created by the sensor contraptions. Restorative reports were produced all the time-dependent on steady checking. The essential server examinations information dependent on the reports created [23].

The Internet of Things was creating continuously and making life simpler for patients and experts. The contraptions, for instance, shrewd meters and wellbeing groups and RFID-based savvy watches and brilliant camcorders help at the same time. Providers should be equipped for dealing with a lot of data and information, which was testing. The capability of IoT for therapeutic offices was accumulated by savvy sensors, which were exact and break down an assortment of health parameters.

4 Research Plan

In this work, an effective early cautioning plan utilizing knowledge capacity was proposed. In this plan, the insight work (Fx) will assume a key job to anticipate the pivotal zone of the system to create early admonitions. For producing this Fx, reasonable strategies will be utilized which will abuse past and current information of the system or node. The proposed plan will be imitated by utilizing various situations to show their presentation regarding the system's lifetime and complete vitality utilization in the system. The general point was to plan an independent plan which can caution a system before any appalling occurrence or mischief. This exploration will contribute a great deal in the healthcare system for sparing it from any harm or misfortune, to give concerned care to the patients just as to improve generally speaking unwavering quality of the healthcare system.

5 Conclusion

Nowadays, health checking is ending up basic as there is an extension in health outline in light of various whimsical infections. In this paper, we have discussed distinctive health observing systems, taking propelled cell phones like a gadget. By using such checking structures, the healthcare specialists can screen, dissect, and counsel their patients from a remote zone continually and master or patient can get the opportunity to report through on the web. It moreover proposes to show the all-around requested headway approach thought in prototyping the Intelligent E-health section including adroit remedial packaging and therapeutic sensor systems. For the explanation behind showing the feasibility of the methodology, and exergaming stage and an affliction organization, mechanical assembly was used as an examination circumstance.

References

1. Abdullah AS, Rajalaxmi RR (2012) A data mining model for predicting the coronary heart disease using random forest classifier. In: Proceedings on international conference on recent trends in computational methods, communication and controls, pp 22–25
2. Alkeshuosh AH, Moghadam MZ, Al Mansoori I, Abdar M (2017) Using PSO algorithm for producing best rules in diagnosis of heart disease. In: Proceedings of international conference on computer and applications (ICCA), pp 306–311
3. Al-milli N (2013) Back propagation neural network for prediction of heart disease. J Theor Appl Inf Technol 56(1):131–135
4. Devi CA, Rajamohana SP, Umamaheswari K, Kiruba R, Karunya K, Deepika R (2018) Analysis of neural networks based heart disease prediction system. In: Proceedings of 11th international conference on human system interaction (HSI), pp 233–239. Gdansk, Poland
5. Anooj PK (2012) Clinical decision support system: risk level prediction of heart disease using weighted fuzzy rules. J King Saud Univ-Comput Inf Sci 24(1): 27–40
6. Baccour L (2018) Amended fused TOPSIS-VIKOR for classification (ATOVIC) applied to some UCI data sets. Expert Syst Appl 99:115–125
7. Diwaker C, Sharma A, Tomar P (2019) IoT's future aspects and environment surrounding IoT. In: Amity international conference on artificial intelligence (AICAI)
8. Jangra A, Mehta N (2018) IoT based early warning model for healthcare system. J Netw Commun Emerg Technol 8(2):78–81
9. Cheng C-A, Chiu H-W (2017) An artificial neural network model for the evaluation of carotid artery stenting prognosis using a national-wide-database. In: Proceedings of 39th annual international conference of the IEEE engineering in medicine and biology society (EMBC), pp 2566–2569
10. Almalchy MT, Popescu N, Algayar SM (2018) A critical analyze on healthcare monitoring systems requirements. In: Proceedings of IEEE international conference on computational science and engineering
11. Begum S, Parveen U (2016) U-Healthcare and IoT. Int J Comput Sci Mob Comput 5(8):138–142
12. Rahmani A-M, Thanigaivelan NK, Gia TN, Granados J, Negash B, Liljeberg P, Tenhunen H (2015) Smart e-health gateway: bringing intelligence to Internet-of-Things based ubiquitous healthcare systems. In: Proceedings of 12th annual IEEE consumer communications and networking conference (CCNC)
13. Esfahani HA, Ghazanfari M (2017) Cardiovascular disease detection using a new ensemble classifier. In: Proceedings of the international conference on knowledge based engineering and innovations (KBEI), pp 1011–1014
14. Ukis V, Rajamani ST, Balachandran B, Friese T (2013) Architecture of cloud-based advanced medical image visualization solution. In: Proceedings of IEEE international conference on cloud computing in emerging markets (CCEM), pp 1–5
15. Rao B (2013) The role of medical data analytics in reducing health fraud and improving clinical and financial outcomes. In: Proceedings of 26th IEEE symposium on computer-based medical systems (CBMS)
16. Ray P (2014) Home health hub internet of things (H3IoT): an architectural framework for monitoring health of elderly people. In: Proceedings of international conference on science engineering and management research (ICSEMR), pp 1– 3
17. Anzanpour A, Rahmani A-M, Liljeberg P, Tenhunen H (2015) Internet of things enabled in-home health monitoring system using early warning score. In: Proceedings of the 5th EAI international conference on wireless mobile communication and healthcare

18. Azimi I, Anzanpour A, Rahmani A-M, Liljeberg P, Salakoski T (2016) Medical warning system based on Internet of Things using fog computing. In: Proceedings of IEEE international workshop on big data and information security (IWBIS)
19. Gope P, Hwang T (2016) BSN-care: a secure IoT-based modern healthcare system using body sensor network. IEEE Sens J 16(5):1368–1376
20. Basanta B, Huang Y-P, Lee T-T (2016) Intuitive IoT-based H2U healthcare system for elderly people. In Proceedings of 13th IEEE international conference on networking, sensing and control. Mexico
21. Plageras AP, Psannis KE, Ishibashi Y: IoT-based surveillance system for ubiquitous healthcare. In Proceedings of 11th IEEE international conference. Greece
22. Ukeil A, Bandyoapdhyay S, Puri C (2016) IoT healthcare analytics: The importance of anomaly detection. In: Proceedings of 30th IEEE international conference on advanced information networking and applications
23. Gupta P, Agrawal D, Chhabra J (2016) IoT based smart healthcare kit. In: Proceedings of IEEE international conference on computational techniques in information and communication technologies

Chapter 12
Artificial Intelligence in Bioinformatics

Vinayak Majhi and Sudip Paul

1 Introduction

The revolution of computer comes in front after the discovery of the semiconductor devices. After the Second World War, application of electronics and computer science took a huge place all over the world. But, the application of computer science in medical sector or more precisely in healthcare science came after a decade later [1]. The application of computer science is to capture and process data with several designed algorithms to extract information from the captured data. Bioinformatics is the field of study where we process the statistical biological data and generate different types of information by using electronics and computer science algorithm [2]. Nowadays, bioinformatics subject itself becomes a multidisciplinary subject with software engineering, electronics, statistics, computer science, mathematics and definitely biology itself [2, 3]. The main aim of bioinformatics is to better understand the disease caused by the genetic disorder and also find the unique adaptation and the difference between different populations. A bioinformatics solution generally comes through four major steps.

- Collection of statistical data from the different biological subjects.
- Designing a computational tools or algorithm based on the feature extracted upon the taken biological subjects.
- Run a simulation program of designed algorithm on the statistical data.
- Finally, testing of the algorithm based on the simulation result.

In this above mentioned steps, artificial intelligence (AI) can take place to generate better performance to find the optimized result based on the statistical data. In the domain of AI, machine learning (ML) already had proven its significance in the application of different domains like speech recognition, biometric data analysis, Internet of Things, natural language processing and many more (Fig. 1).

V. Majhi · S. Paul (✉)
Department of Biomedical Engineering, North-Eastern Hill University, Shillong, India

S. Jain et al. (eds.), *Advances in Computational Intelligence Techniques*,
Algorithms for Intelligent Systems,
https://doi.org/10.1007/978-981-15-2620-6_12

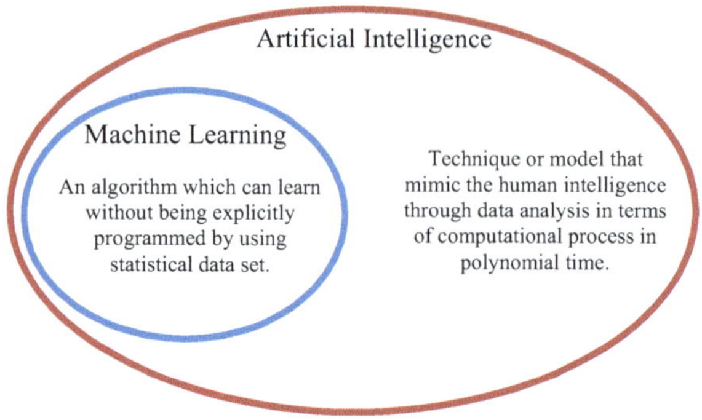

Fig. 1 Machine learning is a subset of artificial intelligence

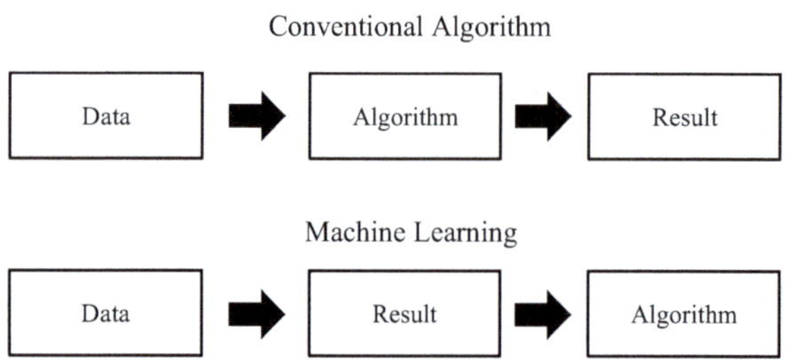

Fig. 2 Difference between convention algorithm and machine learning

In conventional algorithm, we try to find the result by providing the inputs through the designed algorithm after processing. But, in ML, we provide the inputs and corresponding results to find the suitable algorithms which can generate the desired result for an unknown data set in polynomial time; for that reason, training method in ML is very much essential for tuning the ML algorithm. In general, ML helps to find the pattern in a data set in terms of image, voice, text, signals, etc., depending upon the pattern of feature that we want to match or find in an unknown data set (Fig. 2).

Bioinformatics also is getting benefit from this magical tool of AI from the past few years [4]. In the introduction of ML, we can say that ML itself a combination of statistics and computer science where the designed program can learn and update itself from the given statistical data for its training purpose and give the predictive result for new data after training [4]. For that reason, this ML computational tool can be very useful to the bioinformatics.

Beside bioinformatics healthcare informatics is a term in healthcare management where we also focus on preclinical, clinical and post-clinical data like prediction of disease, symptom analysis, determining the accurate stages of disease, beside genetic data, DNA, RNA, 3D structure analysis. In bioinformatics and healthcare informatics, one major issue is the data size that growing exponentially day to day. According to one survey, it was told that in 2012, the healthcare data was approximately 550 PB that will reach 26,000 PB in 2020 [5]. It was found that a tumor genome matching with a normal sample tissue takes nearly 1 TB of uncompressed data; further, it can be reduced 10 times after compression. So, one million genome will take nearly 1000 PB data [6]. Handling this huge amount of data day by day is going to be very difficult. To process this data also increases the cost and time. Here is the role of big data analytics that can capable of extracting the information from the data, and it is also capable of predicting the outcomes of the processed data. Machine learning is one of the common solutions to this big data management. It acts nearly same as data mining which tries to capture the similar pattern in all data set. The machine learning algorithms help to extract data format understandable to the processing program [5].

2 Adaptation of AI in Bioinformatics

Bioinformatics and artificial intelligence correlate with each other from 1960 when first a professor from Stanford named J. Lederberg was interested to create demand of equipment assisted by some program in the field of biology. But then, bioinformatics term was not established [7]. In the beginning of 1970s, Ben Hesper and Paulien Hogeweg initiated to use the term "bioinformatics" for the research they wanted to do, defining it as "the study of informatic processes in biotic systems" [8]. In these days, the concept of artificial intelligence trying to make its establishment by giving the solution to the complex real-world problem that we called in generally NP problem, which cannot be solved in polynomial time with deterministic algorithms. DENDRAL was the first project based on the task specific knowledge-based problem that can capable of recognizing the organic compound from using as a source of heuristic and automatic knowledge acquisition from the mass spectrometric database [9]. After twenty years, during PROTEAN project, the first three-dimensional protein structures were determined using high-resolution nuclear magnetic resonance. Here, also some knowledge-based system was successfully developed that can capable of interpreting the crystallographic data [10]. The Meta-DENDRAL project was the actual initiation of the machine learning approach in realistic application [11]. Bioinformatics can be defined as a combined task of analyzing biological signals and interpreting those signals to get some valid predictive result and also managing the biological data management. For gathering biological sequencing data with its structure, European Molecular Biology Laboratory (EMBL) with GenBank is established. The goal is to match different kinds of new DNA sequence with available DNA sequencing database. These progresses lead us to the current research technique to encompass different functional data sets of gene or protein with its structural analysis and also the metabolic rout comparison that can be done for different kinds of individual species during the occurrence of clinical or preclinical trials [12]. Now, AI can be found in all parts of the bioinformatics [7]. As because bioinformatics dales with different types of NP-Hard problems which are great challenge for Artificial Intelligence. Besides bioinformatics, AI is being applied in healthcare informatics management. Now, AI is applied in digital medical image acquisition and to find the problem sector from that image. Symptomatic analysis of patient data now can predict the type of disease with its stage by analyzing the previous statistical data. Machine learning and big data analytics are now the main key feature of the different AI tools for heathcare informatics.

Bioinformatics actually provides the pathway to gather the large amount of molecular biological data as well as the way of storing and processing those data. Besides this, DNA array can be generated, and it will also help to find the location of the areas in protein coding [2]. How AI analyze the proteomic data and genetic expression by reverse engineering of gene network using correlation identification, clustering and self-organizing maps is difficult to understand, but it can capable of measuring thousands of data set in one microarray whichis a very large data set which needs extended biological knowledge of human expertise to find a simple relation of between two gene expression [13]. AI is probable not suitable for those problem where finding the best solution; but it works very efficiently where to determine which is the better mean, which one is weak and which one is the strong in certain defined constrain.

DNA chip made up of thousands of nucleotide fragments; this conception is brought to light by the multiple gene expression measurement. Only this method is used to identify the gene coding DNA from the sample and spot the subsets of genes [14]. In genomics and proteomics rare, feats are more advantageous in data mining process. The introduction of novel drugs and its effectiveness increase the success rate in identification of gene-related disorders. Mathematics and statistics can deal with the computation and interpretation problems associated with the molecular level data. Some convenient computational tools are designed by the help of computer science and information technology whereas a post-processing tools AI can be enforced to get the best result [15]. Some different applications of the bioinformatics are as follows:

2.1 Sequence Analysis

All genes of its genome determine the genetic basis of any organism. Sequence analysis is a process which is very commonly used to learn its structure, function, features [16]. Process of sequence analysis has various types in the field of bioinformatics like

- Pair-wise Sequence Alignment
- Multiple Sequence Alignment
- Pattern Search
- Construction of Phylogenetic Trees.

In pair-wise sequence alignment, we analyzed the similarities and the differences of DNA or protein sequence in between two pairs [17] (Fig. 3).

Multiple sequence alignment (MSA) is problem defined as to find the technique to align DNA or protein sequence as a number of three or more. Through this, we can find the subregion or the embedded pattern of the taken sequences.

Fig. 3 Pair-wise sequence analysis [3]

```
AGGCTAATCAGATTAGGCAGCCTTAAGCCATA
CAGGCTGATCAGATTGGCTAGCTTGCAAGATAGAAC

-AGGCTAATCAGATTAGGCAGCCTT--AAGCCATA----
CAGGCTGATCAGATT-GGCTAGCTTGCAAG--ATAGAAC
```

In the problem of sequence analysis, sometimes it is required to find a small sequence in a large sequence database. Here, several pattern-based searches are required to find a sequence in thousands of the gene sequence [3].

Phylogenetic analysis is the determination of the evolutionary hierarchy between species. By this analysis, we can determine the origin of a species. It is also used in the study of biodiversity [3].

In sequence analysis, AI can be used to get the most optimum solution. There are several AI algorithms that can be used for sequence analysis like hidden Markov model (HMM). The profile HMM is one of the popular HMM models in molecular biology [18]. For behavioral phenomenon of DNA observation in a living cell fuzzy, it also can predict the basic molecular behavior [19]. Artificial neural network is also used for DNA sequence analysis [20]

2.2 Protein Structure Prediction

The three-dimensional structure of protein from its amino acid can be predicted. The determination of the secondary, tertiary or quaternary structures of proteins is very difficult. To solve this problem, either the method of crystallography or bioinformatics tools is used [21]. The 3D structure of the protein is determined by the high-resolution 2D nuclear magnetic resonance (NMR). To predict the various shapes or structures, AI pattern recognition tools become very helpful. In bioinformatics, the pattern recognition method is used to determine the 3D structure of protein. TEXTAL computer program can assume automatic structure of protein by X-ray crystallography through the measurement of electron density mapping [22, 23].

2.3 Genome Annotation

Genome annotation can be used to understand the regulatory sequence and protein coding. This method is used to identify the locations of all genes, coding regions and structure of genome [24]. The main function of the genome annotation is to find all types of functional elements in a gene; also, it finds the number of occurrences [25]. Profile hidden Markov model and position weight matrix are very useful for this purpose [25].

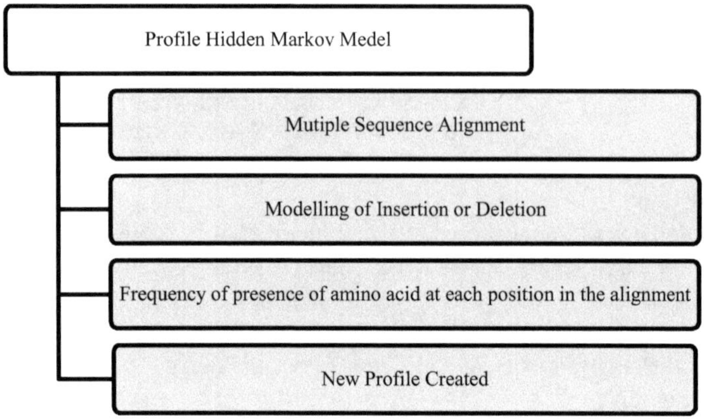

Fig. 4 Steps of profile HMM [26]

HMM was mainly used in speech recognition, and it was showing promising results in it through the past decades. It is also very useful in solving molecular sequencing problem analysis. The profile HMM is helpful to reconstruct the profile from the previous traditional profile. The basic advantages of profile HMM over conventional profile are for creation of profile based on the probabilistic approach which is basically a heuristic method that deals with the theory behind gap and insertion scores. The profile HMM can be derived from 10 to 20 aligned sequences where a standard profile required 40–50 aligned sequences. So, we can conclude that the HMM helps to create good profile in lesser effort and skills [26] (Fig. 4).

2.4 Comparative Genomics Hybridization

To establish the genomic structure and behavior between different kinds of biological species, the genomic features such as gene, CDS, mRNA, tRNA and rRNA will be measured. To detect the whole path of evolution, the inter-genomic maps are used by the researchers. The related data about the point mutations and large chromosomal segments can be found by these maps [27]. In comparative genomic hybridization (CGH), we check the imbalance of chromosome. Self-organizing map (SOM) can be done by artificial neural network, where the unsupervised learning is used for clustering the input data with similar vector in the same cluster. Kohonen's SOM is the most useful for analyzing the CGH [28].

2.5 Drug Discovery

The bioinformatics tools are used in drug discovery based on molecular disease. The researchers invent medicines and drugs which can suit more than 500 genes depending on the disease and diagnosis management. The delivery of a particular drug depends on the cells to be targeted, and multiple computational tools can be helpful for this purpose [29]. Artificial intelligence is used in genomic drug discovery and genomic medicine development. Molecular dynamic simulation is being done by AI for binding affinity prediction to produce mutated proteins and drug [30].

In 1960, quantitative structure–activity relationships (QSAR) were introduced as alternative drug discovery purpose. The idea behind QSAR is to use the known response, that is, the activity of a simple compound to predict the response or activity of an unknown complex compound which is designed as a different combination of the basic compound module. This QSAR system is basically based on structural, quantum mechanical and other descriptors like electronic, hydrophobic and steric attributes of a molecule. In further, QSAR adopt several other descriptors to design the molecular compound like molar refraction index (MR), shape of the compound, quantum chemical descriptor, etc. With the increasing parameter, it becomes difficult to establish the correlation between the parameters either the help of artificial intelligence. AI basically helps to establish the correlations between all the parameters adopted by the QSAR. Sometimes, the target protein structure may remain unknown; in that case, the drug target may be analyzed by the experimental method which is called pharmacophores identified feature, and this includes the 3D QSAR technique. QSAR uses the pattern recognition, clustering, feature selection, classification features of the AI to resolve the complexity of its own process in the way of drug discovery [31]. Besides this, several AI tools like artificial neural network, fuzzy modeling, support vector machine are being used for this purpose, but till dated, there are no such drugs approved by Food and Drug Administration (FDA) which fully inspired by the artificial intelligence [31, 32].

2.6 Healthcare Application

AI also has a huge impact on healthcare application. Many diagnosis tools have been governed by the AI for the detection of different abnormalities. The International Medical Device Regulators Forum (IMDRF) introduced "Software as a Medical Device (SaMD)" where software only be used for some medical purposes without having any hardware dependency [33]. Under the Federal Food, Drug, and Cosmetic Act (FD&C Act), US Food and Drug Administration also considers some medical purposes which have its required action on diagnosing treating, mitigation prevention and other purposes also [33]. But, there are so many rules and regulations also implemented with regular upgradation for safety purpose. Figure 5 shows the total

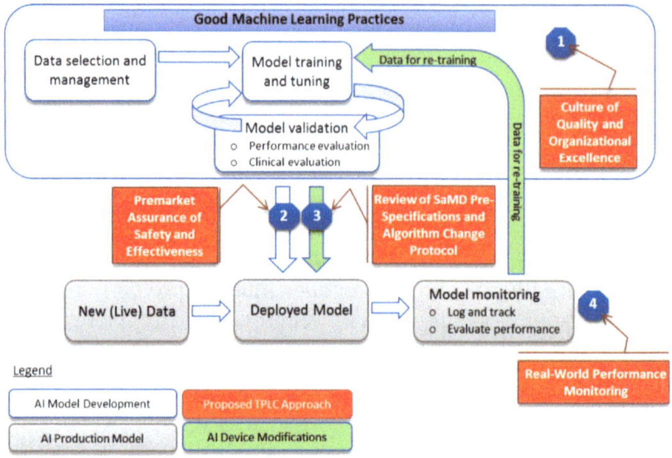

Fig. 5 Overview of FDA's TPLC approach on AI/ML workflow [33]

product lifecycle (TPLC) approach of evaluating a medical software product from premarket development to post-market performance.

3 Tool of Bioinformatics

Bioinformatics tools are the amalgamation electronics and computer engineering in the field of bioinformatics. The preliminary target of these tools is to supply the data about portrayal of qualities and phylogenetic examinations. These tools also find out auxiliary and physiochemical properties of proteins. This information is useful to think about the behavior of biomolecule in a living cell. All these bioinformatics tools are designed for some specific tasks to perform. Several AI-based algorithms also used in the main core analyzing sector of these tools. The goal of the bioinformatics tools is to provide error-free information of biomolecule, gene and protein. It also helps to manage the data bank for large genomic data. Various types of bioinformatics tools used for distinctive applications are as follows.

3.1 Gene Identification and Sequence Analyses

Grouping investigations can help to find out the comprehension and highlights of a biomolecule like nucleic acid or protein. This process starts through sequence recovery of relevant molecules from open databases. Depending on the requirement, various tools are discovered to empower its competence and focus the findings such as capacity, structure and homologues with an incredible precision.

To compare the sequences of proteins, nucleotides, DNA and RNA, basic local alignment search tool (BLAST) is used which is a very efficient searching tool. It also helps to find the regions of local similarity between nucleotide sequences, and protein sequences can be compared using this application. BLAST also helps to find the similar member of the gene families. For unknown species identification, BLAST is very much helpful, and it can identify correct or similar homologues species by analyzing the DNA sequence. It is also capable for generating the phylogenetic tree. It also uses the DNA mapping of known chromosome. For known species gene sequencing, BLAST helps to identify the position of the query sequence of chromosome location. It can also be used to map annotations from one organism to another. It also identifies the common gene expression in two related species [34].

"HMMER" is another free software used for identification and analysis of homologous protein sequences from the individual databases. It used the profile HMM for sequencing the protein. HAMMER also helps in DNA/RNA queries upon the databases of DNA/RNA sequence, search nucleotide sequence in nucleotide profile [35].

"Clustal" is a very popular tool for multiple sequence alignment. There are several versions of Clustal which is available since 1988, but Clustal Omega is the most promising and standard version. It gives average result for small sequencing. But, it will work fine when large number of data set is given. This is also capable of more than one million sequences at a time in single process [36].

"Sequerome" is a Web-based sequence profiling tools. It is basically a Java-based program that generates query for BLAST. Sequerome works in front-end API, and BLAST works in back-end [37].

"ProtParam" is computational tools which deduct the physicochemical properties from protein sequence [38].

Gene production using multiple sources of evidence "JIGSAW" is used to identify the gene models through sequence alignment. It is developed in C++, and it can support for near about all species. The nucleotide precision accuracy of this tool is 98%, and nucleotide sensitivity is 90% [39].

Open reading frame finder or ORF finder is used for various bioinformatics analyses like open reading frame in a given DNA sequence. It also produces graphical viewing and data management. It is very much helpful for complete and accurate sequence submission [40].

To upstream the quality of promoter sequences, prokaryotic promoter prediction tool (PPPT) is being used [41].

"Virtual Footprint" is an online platform which gives the accurate and promising results for prediction of transcription factor binding sites (TFBSs). It provides the complex DNA pattern generated from the bacterial genome [42].

The intrinsic transcription terminators identified in bacterial genome sequences and plasmids will be stored in "WebGeSTer DB" database. It is the largest database which is freely available of intrinsic transcription terminators [43].

In order to identify the complete gene structure in genomic DNA, "GENSCAN" program will be used. It can predict the gene location with exon–intron boundaries [44].

The tool "Soft berry" is a cloud computing service, mainly used for alignment and genome comparison and regulation analysis of genome including plants and animals [45].

3.2 Phylogenetic Analysis

The genetic relatedness is depending on the transformative relationship among the group of same molecules, organisms and life forms. It is used to find the evolutionary bonding between organisms. After the phylogenetic analysis, it generates a tree structure called cladograms to find the ancestral sequence [46, 47].

Molecular evolutionary genetics analysis "MEGA" is a user-friendly and sophisticated software used to analyze the protein sequence data from classes and inhabitants. It is developed based on phylogenetic trees of DNA and protein sequences in the year of 1990 to work on the transformative closeness [48].

"MOLPHY" is a program package used to examine the subatomic phylogenetic based on the probability technique.

Phylogenetic analysis by maximum likelihood "PAML" and phylogeny inference package "PHYLIP" utilize the probability technique for phylogenetic analysis, and each package contains several computational tools.

The open-source Web-based libraries such as BioJs tree and Treeview are available which contain phylogenetic tree analysis.

"Jalview" is the software used to edit and analyze the multiple sequence alignments [45].

3.3 Predicting Protein Structure and Function

In preliminary stage, the protein molecules look like unorganized amino acid strings. These molecules achieve three-dimensional (3D) structures as biological dynamics at the ultimate stage. The action of the biological functionalities depends on breakdown of protein into a right topology which is pre-imperative. Now, it can be concluded that the 3D structure of protein is essential for its feature extraction, and by X-ray crystallography or nucleic magnetic resonance, these kinds of figure can be observed. Thermodynamic equilibrium physiochemical standards help to evaluate the comparison of forecasting structures and validity of simulation algorithm with a global marginal free energy of protein surface. The various tools used to identify the protein structure and function are listed in Table 1.

Table 1 Protein structure and function identification tools [45]

Name of the tool	Applications
CATH	It is a self-loader tool to arrange the association of proteins
RAPTORX	To forecast protein structure depending on either multi-layout threading or solitary
JPRED and APSSP2	Used to predict the auxiliary structures of proteins
PHD	To predict the neural network structure
HMMSTR	To forecast the arrangement structure relationships in proteins
MODELLER	To predict the 3D structure of protein based on comparative modeling
PHYRE and PHYRE2	These are the Web-based servers used to predict the protein structure

Table 2 List of sequence database

Type of database	Database name
Genome databases	Ensembl, PIR
Nucleotide databases	DNA Data Bank of Japan, European Nucleotide Archive, GenBank, Rfam
Protein databases	Uniprot, Protein Data Bank, Prosite, Pfam, SWISS PROT, Proteomics Identifications Database, InterPro
Signaling and metabolic pathway databases	HMDB, KEGG, CMAP, PID, SGMP
Miscellaneous databases	Medherb, Reactome, TextPresso, TAIR, dictyBase

3.4 Genome Sequence Databases

In NCBI GenBank, there are collections of genomes near about 2.5 million different species. Each gene sequence carries information about the literature, bibliography, organism and including some other features like promoters, terminators, exons, introns, coding regions, repeat regions, untranslated regions and translations [45].

Besides this, there are some other databases also available by its feature and content (Table 2).

3.5 Drug Designing

In the earlier stage, when the modern bioinformatics tools were not invented, to discover the new compound in the world, the researchers, who belong to different fields like pharmacology, clinical sciences and chemistry, worked together for introducing the new compound to the world. The above methods come to end by the invention of the bioinformatics, and revelation and planning in the medical domain are encouraged. It is not difficult to impart the molecules using the software compared to other different approaches. Nowadays, it is possible to design the highly effective drugs by the help of computer-based simulating software which is called as computer-aided drug design (CADD). The drug development method is much complicated and difficult. There are four basic steps to design a new drug [49, 50].

- Drug Target Identification
- Target Validation
- Identification of the Lead
- Optimization of the Lead.

The popular drug databases developed by various organizations to facilitate the researchers and scientists who are interested to design the new drug are as follows

- Potential Drug Target Database (PDTD)
- Therapeutic Target Database (TTD)
- Target Database (TDR)
- Manually Annotated Targets and Drugs Online Resource (MATADOR)
- Drug Bank
- Tropical Disease Research
- Tuberculosis (TB) Drug Target Database.

The various molecular dynamic tools such as Abalone, Ascalaph, Discovery studio, Amber and FoldX are used to get the information related to ion transport, conformational changes of proteins, nucleic acids and complexes occurring in biological system [45].

4 Conclusion

Development in the field of bioinformatics had been done from the few decades. But in the past ten years, with the advancement of computing device and processor, several artificial intelligence programs run very efficiently. In case of database management, now, it is possible to store more information and samples in digital platform. The cloud computing is also helping to the user to use all applications and database remotely. Previously, machine learning algorithms need huge time to train and process. But, with the help of newly developed graphical processor, the deep learning can replace the machine learning method. Now, we can process millions of genome sequencing data within a single process within few hours with most accuracy than the past decades. As if till now, there are no such FDA approved drugs which are fully inspired by the AI. But very few AI can be capable of designing the genome information based customized drug for the person's disease. Till now, AI is also helping to the virtual synthesis of new drug. And the accuracy level of that synthesis is very firmly differing from the actual synthesis. Human genome is a complex case of understanding it's all attributes by its function. The further contribution in the field of bioinformatics can solve more human queries with the help of artificial intelligence in near future.

References

1. Mantas J (2016) Biomedical and health informatics education—the IMIA Years. Yearb Med Inform (Suppl. 1):S92–S102
2. Thampi SM (2009) Introduction to bioinformatics. arXiv:0911.4230
3. Can T (2014) Introduction to bioinformatics. In: Yousef M, Allmer J (eds) miRNomics: microRNA biology and computational analysis. Humana Press, Totowa, NJ, pp 51–71
4. Sree Divya K, Bhargavi P, Singaraju J (2018) Machine learning algorithms in big data analytics. Int J Comput Sci Eng 6:63–70
5. Kumar S, Singh M (2019) Big data analytics for healthcare industry: impact, applications, and tools. Big Data Min Anal 2(1):48–57
6. O'Driscoll A, Daugelaite J, Sleator RD (2013) 'Big data', Hadoop and cloud computing in genomics. J Biomed Inform 46(5):774–781
7. Nicolas J (27 July 2018) Artificial intelligence and bioinformatics. Available from https://hal.inria.fr/hal-01850570
8. Hogeweg P (2011) The roots of bioinformatics in theoretical biology. PLoS Comput Biol 7(3):e1002021
9. Lindsay RK et al (1993) DENDRAL: a case study of the first expert system for scientific hypothesis formation. Artif Intell 61(2):209–261
10. Hayes-Roth B et al (1986) PROTEAN: deriving protein structure from constraints. American Association for Artificial Intelligence
11. Feigenbaum EA, Buchanan BG (1993) DENDRAL and Meta-DENDRAL: roots of knowledge systems and expert system applications. Artif Intell 59(1):233–240
12. Müller UR, Nicolau DV (2005) Microarray technology and its applications. Springer, Berlin
13. Narayanan A, Keedwell E, Olsson B (2002) Artificial intelligence techniques for bioinformatics. App Bioinform 1:191–222
14. Oyelade O et al (2015) Bioinformatics, healthcare informatics and analytics: an imperative for improved healthcare system. Int J Appl Inf Syst 8(5):1–6
15. Bagga PS (2012) Development of an undergraduate bioinformatics degree program at a liberal arts college. Yale J Biol Med 85(3):309–321
16. Koonin EV, Galperin M (2013) Sequence—evolution—function: computational approaches in comparative genomics. Springer Science & Business Media, New York
17. Zvelebil MJ, Baum JO (2007) Understanding bioinformatics. Garland Science, New York
18. Krogh A (1998) An introduction to hidden Markov models for biological sequences. In: Salzberg SL, Searls DB, Kasif S (eds) New comprehensive biochemistry, Chapter 4. Elsevier, Amsterdam, pp 45–63
19. Ji S (2004) Molecular information theory: solving the mysteries of DNA. In: Ciobanu G, Rozenberg G (eds) Modelling in molecular biology. Springer, Berlin, Heidelberg, pp 141–150
20. Ezziane Z (2006) Applications of artificial intelligence in bioinformatics: a review. Expert Syst Appl 30(1):2–10
21. Breda A et al (2007) Protein structure, modelling and applications. National Center for Biotechnology Information (US)
22. Gopal K et al (2006) TEXTAL: crystallographic protein model building using AI and pattern recognition. AI Mag 27(3):15
23. Holbrook SR, Muskal SM, Kim S-H (1993) Predicting protein structural features with artificial neural networks. In: Artificial intelligence and molecular biology. AAAI Press, Menlo Park, p 162

24. Kellis M et al (2014) Defining functional DNA elements in the human genome. Proc Natl Acad Sci USA 111(17):6131–6138
25. Yip KY, Cheng C, Gerstein M (2013) Machine learning and genome annotation: a match meant to be? Genome Biol 14(5):205
26. Eddy SR (1998) Profile hidden Markov models. Bioinform (Oxford, England) 14(9):755–763
27. Pellegrini M et al (1999) Assigning protein functions by comparative genome analysis: protein phylogenetic profiles. Proc Natl Acad Sci USA 96(8):4285–4288
28. Mattfeldt T et al (2001) Cluster analysis of comparative genomic hybridization (CGH) data using self-organizing maps: application to prostate carcinomas. Anal Cell Pathol: J Eur Soc Anal Cell Pathol 23(1):29–37
29. Gill SK et al (2016) Emerging role of bioinformatics tools and software in evolution of clinical research. Perspect Clin Res 7(3):115–122
30. Fujiwara T, Kamada M, Okuno Y (2018) Artificial intelligence in drug discovery. Gan To Kagaku Ryoho 45(4):593–596
31. Duch W, Swaminathan K, Meller J (2007) Artificial intelligence approaches for rational drug design and discovery. Curr Pharm Des 13(14):1497–1508
32. Agrawal P (2018) Artificial intelligence in drug discovery and development. J Pharmacovigil 6:2
33. Food and Drug Administration (2019) Proposed regulatory framework for modifications to artificial intelligence/machine learning (AI/ML)-based Software as a Medical Device (SaMD). Discussion paper and request for feedback
34. Altschul SF et al (1990) Basic local alignment search tool. J Mol Biol 215(3):403–410
35. Krogh A et al (1994) Hidden Markov models in computational biology: applications to protein modeling. J Mol Biol 235(5):1501–1531
36. Daugelaite J, Driscoll AO', Sleator RD (2013) An overview of multiple sequence alignments and cloud computing in bioinformatics. ISRN Biomath. 2013:14
37. American Association for the Advancement of Science (2005) Tools: a bigger blast. Sci 309(5743):1971
38. Garg VK et al (2016) MFPPI—Multi FASTA ProtParam Interface. Bioinformation 12(2):74–77
39. Allen JE et al (2006) JIGSAW, GeneZilla, and GlimmerHMM: puzzling out the features of human genes in the ENCODE regions. Genome Biol 7(Suppl 1):S9.1–S9.13
40. Rombel IT et al (2002) ORF-FINDER: a vector for high-throughput gene identification. Gene 282(1–2):33–41
41. Kanhere A, Bansal M (2005) A novel method for prokaryotic promoter prediction based on DNA stability. BMC Bioinform 6:1
42. Munch R et al (2005) Virtual footprint and PRODORIC: an integrative framework for regulon prediction in prokaryotes. Bioinform 21(22):4187–4189
43. Mitra A et al (2011) WebGeSTer DB—a transcription terminator database. Nucleic Acids Res. 39(Database issue):D129–D135
44. Burge C, Karlin S (1997) Prediction of complete gene structures in human genomic DNA11. Edited by F. E. Cohen. J Mol Biol 268(1):78–94
45. Mehmood MA, Sehar U, Ahmad N (2014) Use of bioinformatics tools in different spheres of life sciences. J. Data Min. Genomics & Proteomics 5(2):1
46. Kidd KK, Sgaramella-Zonta LA (1971) Phylogenetic analysis: concepts and methods. Am J Hum Genet 23(3):235–252
47. Doroshkov AV et al (2019) The evolution of gene regulatory networks controlling Arabidopsis thaliana L. trichome development. BMC Plant Biol 19(Suppl. 1):53
48. Kumar S et al (2008) MEGA: a biologist-centric software for evolutionary analysis of DNA and protein sequences. Brief Bioinform 9(4):299–306
49. Katara P (2013) Role of bioinformatics and pharmacogenomics in drug discovery and development process. Netw Model Anal Health Inform Bioinform 2(4):225–230
50. Majhi V, Paul S, Jain R (2019) Bioinformatics for healthcare applications. In: 2019 Amity international conference on artificial intelligence (AICAI)

Chapter 13
English Numerals Recognition System Using Novel Curve Coding

Binod Kumar Prasad

1 Introduction

The domain of the presented work is offline optical character recognition (OCR). The old documents written on papers are not safe for long time because of aging effect of papers. All these documents need to be typed and stored as soft copy in computers. Global estimation tells that about 250 billion USD is spent per year to key information from 1% of the available documents which makes the manual process of keying, very costly. Again, manual typing is time-consuming and very prone to errors. Optical character recognition provides with a ready solution to this menace. It enables a machine to recognize any text automatically. OCR is a technique that transfers the printed documents to a computer-readable form. It helps automated archiving and retrieving of old documents, especially old books. Nowadays, OCR finds a wide range of applications for mail sorting, bank cheque processing, reading machines for the blind, automatic number plate recognition in addition to handwritten text recognition. OCR applies the technique of machine learning (ML). ML has been a dynamic field with rapid evolutions aimed at developing and improving algorithms that allow computers to learn from data and recognize patterns. Offline numeral recognition system consists of the process of classification among ten different classes. However, none of the reported works related to handwritten numeral recognition are focused on how to tackle the problem arising out of the two very frequent writing formats of English numerals '4' and '7' as shown in the problem statement and motivation section. This paper incorporates Multiple Hidden Markov Model to address format variations where the two different formats of the numerals have been considered as two different classes.

The flexibility and hence reliability of an optical character recognition system cannot be enhanced remarkably unless the local curves are encoded efficiently to find

B. K. Prasad (✉)
Guru Nanak Institutions Technical Campus, Ibrahimpatnam, Hyderabad, Telangana, India

S. Jain et al. (eds.), *Advances in Computational Intelligence Techniques*,
Algorithms for Intelligent Systems,
https://doi.org/10.1007/978-981-15-2620-6_13

more genuine local feature elements out of it. The current paper introduces differential approach in terms of distance and slope to encode the local curves to address structural variations in handwritten numeral samples. MHMM performs classification, and the results are post-processed with SVM to get final outcomes. The paper organization is as follows.

Section 2 reviews the literature; Sect. 3 describes the problem; Sect. 4 details out proposed multiple HMM model; Sect. 5 deals with preprocessing; Sect. 6 contains methods of feature extraction; Sect. 7 presents experiments and their results followed by conclusion in the last section.

2 Review of Related Works

The individuality of one's handwriting adds to the complexity of its recognition. Literature reports various techniques to tackle it [1]. Karungaru et al. [2] have incorporated neural network into classify characters. Features have been derived from star-layered histograms developed from intersections of lines through center of gravity up to the contour and the character. The system gets an overall recognition rate of 97.1% on MNIST database. Qacimy et al. [3] have differentiated the classes of digits by means of three types of DCT coefficients. Classification with SVM on MNIST dataset yielded accuracy of 98.66%, 98.71%, 98.73%, and 98.76% respectively. Authors [4] have reported a cascaded recognition system to recognize mixed digits of Devanagari and Bangla under Indian script and English. They have used wavelet-based multiresolution representations and multilayer perceptron (MLP) to provide misclassification rates of 0.81% and 1.01% on training and test MNIST English samples, respectively, with overall accuracy of 98.79%. Paper [5] has presented a character recognition system for cursive English handwriting in a medical prescription to read the names of medicines. It uses horizontal projection method for text-line segmentation and vertical projection histogram method for word segmentation. Feature extraction has been done by means of convex hull algorithm, whereas classification has been done with SVM. The system is reported to have an accuracy of 85%. Paper [6] has put forward a study regarding the effect of depth of convolutional neural network (CNN) architecture on recognition accuracy of handwritten Devanagari characters. It has been concluded that CNN-BLSTM hybrid architecture has dominated over state-of-the-art Devanagari character recognition. Authors [7] have used histogram of oriented gradient for feature selection and SVM as classifier. It produced an accuracy of 98.05% on CMATERDB dataset. The authors [8] have proposed an integrated system of OCR and text-to-speech conversion to help visually impaired people to read a document. The system uses OCR and text-to-speech module of mobile phones of new generation. Authors [9] have proposed orthogonal feature detectors with ANN for improvement in the performance in noisy conditions. They have reported 56.4% relative improvement in the rate of recognition on MNIST database over conventional method of learning without orthogonalization. Authors [10] have achieved recognition rate of 94.76% on MNIST test database by means of scaling and rotation invariant system using structural characteristics as global features and local descriptor based on scale-invariant feature transform (SIFT). LeCun et al. [11] have reviewed several methods of HCR and found CNN with gradient-based learning method to be superior as it can alleviate variability of 2D shapes. GTN also uses CNN along with global training techniques to yield promising results. Authors [12] have combined conditional random fields with deep learning to recognize handwritten words. Deep features are learned, and sequences are labeled in a unified framework. Deep structure is pre-trained with restricted Boltzmann machines, and optimization of the entire network has been done with online learning algorithm. Authors [13] have proposed CNN and SVM-based hybrid classifier to recognize handwritten digits where CNN extracts features followed by SVM as classifier. In [14], authors have extracted features from signature of characters and applied Optimum path forest classifier for recognition of digits. They have got satisfactory results with, respectively, lesser time of processing. Agarwal and Goswami [15] have reported a robust and simple approach of text recognition in vehicle plate based on edge detection and segmentation. The whole work of recognition has been done in three steps. First license plate is recognized, and then characters are segmented followed by their recognition. In [16], authors have extracted seven sets of features and then made different hybrid combinations out of them. Classification with artificial neural network using backpropagation algorithm has produced a correct rate of recognition of 99.25% on MNIST database. Wu et al. [17] have put forward a trial to identify a writer independent of handwritten text. Handwriting image is first segmented into word regions by means of Laplacian of Gaussian (LoG) mask. Scale-invariant feature transforms (SIFT) along with scales and orientations of the word regions are determined and used to prepare scale and orientation histogram (SOH) in the enrollment phase. These SIFT descriptors and SOH help the identification of writer. Wang et al. [18] have proposed Li-KNN algorithm to recognize handwritten numerals in MNIST dataset. It has shown an average error rate of 21.02 $\pm$ 0.98 which is lesser than that with LieMeans and LieFisher methods of classification. Authors [19] have resorted to newly developed novel unbalanced decision tree architecture, having binary SVM as its unit has been harnessed to recognize handwritten digits. Elleuch et al. [20] have claimed that SVM classifier has not been used as frequently for Arabic handwriting recognition as the classifiers like ANN and HMM. The paper has examined Arabic handwriting recognition using SVM. Handcrafted feature is incorporated into multiclass SVM with RBF kernel. The proposed system is tested on handwritten Arabic character database (HACDB) to give satisfactory results. Authors [21] propose SVM-based handwriting recognition system which incorporates Freeman chain code along with transitional information in vertical and horizontal directions of the Persian numeral images to extract features. The feature extraction technique has been claimed to be free from normalization of images. Authors [22] have put up a successful trial to decrease the complexity of using SVM in one-against-rest mode. They have developed a distribution for the negative samples and the 'hybrid' classifiers determine a hyperplane between positive samples and this probability distribution. The logic estimates the distribution of the background only once, and then the same model trains the classifiers for all visual classes. Thus, it helps significantly reducing the training complexity. A mirror image reflection method has been brought forward [23] to read document backside layer characters by using neural fuzzy hybrid system. It will also facilitate reading the old documents written on palm leaves by our ancestors which are not readable from the front side due to aging effect. The backside-layer input characters are converted into neural inputs and transformed it into fuzzy sets wherefrom the related output responses are generated. The paper [24] proposes local directional pattern and gradient directional pattern-based process for feature extraction followed by k-nearest neighbor (KNN) and support vector machine (SVM) classifiers to read Bangla numerals. It reports an accuracy of 95.62% on the benchmark dataset (CMATERdb 3.1.1). Authors [25] claim a record 97.4% accuracy in recognizing Arabic numerals by dint of a novel algorithm using deep learning neural networks and appropriate activation function. A 3D handwriting character recognition system has been presented [26] based on a symbolic representation of angular velocity signal generated from gyroscope sensor in a smart phone. Experiments have been executed on a real database of the 26 lowercase letters in English alphabet.

3 Problem Statement and Motivation

Individuality of handwriting poses a new threat to numeral recognition. It reflects in the form of either structure or format of the handwritten samples of digits, thereby giving rise to structural and format variations.

3.1 Structural Variations

People do not care for minor structural changes strictly while writing. Table 1 shows structures of some digits in standard (printed) form and handwritten form. It is evident from the table that the encircled part of the digits '2', '4', '5', and '7' has been written curved while the same part in the respective printed form is flat or linear. On contrary, the encircled part of handwritten digits '3' and '9' has been written flat which was supposed to be curved as in the corresponding printed form. As a consequence, handwritten digits like '3' or '9' show linearity and '2' '4' '5' and '7' are found to show considerable amount of curvature. A straight line is also considered as a curve with very long or infinite radius. Therefore, curves are to be analyzed properly in order to establish a promising numerals classification system. This desired criteria motivated the current work to rely upon features representing a curve very accurately. Such features are helpful in increasing flexibility of the system. The system, when trained with correct and deformed handwritten samples, enables it to assimilate minor differences, and the probability of a testing sample getting misclassified easily becomes very low.

Table 1 Structure of printed and handwritten digits

Printed digit	Handwritten digit (s)
2	
3	
4	
5	
7	
9	

3.2 *Format Variations*

In addition to the problem due to the linearity and curvature, different styles of writing of some of the numerals challenge the recognition efficiency of a system. The digits '4' and '7' are often written in the following formats [Fig. 1].

The paper tackles structural variations by means of *novel* distance and slope codings (Sect. 6), whereas format variations have been tackled with Multiple Hidden Markov Model (MHMM) of classification.

4 Proposed Multiple Hidden Markov Model (MHMM)

MHMM makes two different HMM models termed as 'HMM 4a' and 'HMM 4b' for digit '4', whereas 'HMM 7a' and 'HMM 7b' for digit '7' for the two different patterns shown in Fig. 1. The comparator yields the result of the classifier. The outcomes in favor of both the subclasses are considered as the outcome of the corresponding class. The scheme of MHMM has been shown in Fig. 2.

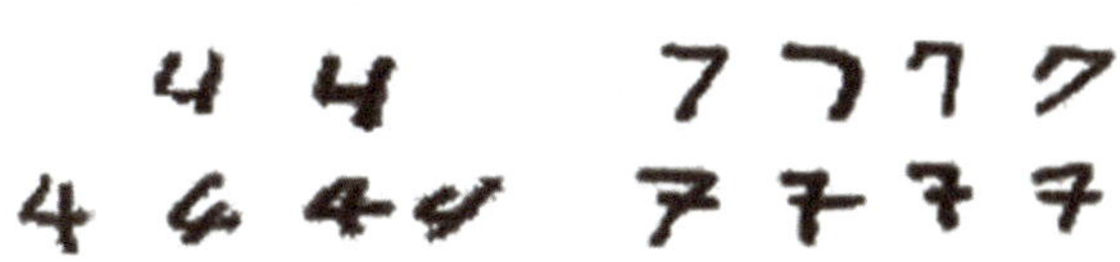

Fig. 1 Two different patterns of numerals '4' and '7'

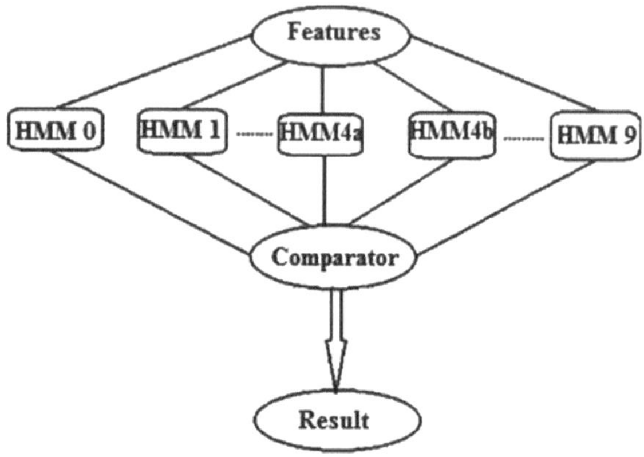

Fig. 2 Proposed multiple HMM module

5 Preprocessing

The paper deals with both CENPARMI and MNIST training and test data. The samples of both these data are resized to a common size of 50 × 50 keeping the aspect ratio intact so that the same MATLAB code for feature extraction and classification could be applied to both these data. Due to resizing the images to 50 × 50, more number of sample points are obtained while performing delta distance coding (Sect. 6), thereby representing the local curves more genuinely without increasing the system complexity. Strokes play a vital role in identification of samples. Stroke is the direction of writing of a part of a character sample. Raw samples are preprocessed to make strokes clear. Stray marks in sample image space give rise to false pixels or isolated pixels that produce erroneous feature elements, and hence, learning and recognition processes of classifier are affected badly. Median filter has been used in the proposed system (Fig. 3) to avoid such probable noisy pixels without any edge blurring effect. Thresholding has been done by means of Otsu's method that binarizes the gray-level images. Skeletonization is done to obtain single-pixel thickness of the numeral images. It is preceded by thickening so that no part of an image could disappear because of thinning operation of skeletonization. Thickening includes the process of dilation which bridges any unintentional discontinuity (gap) while writing

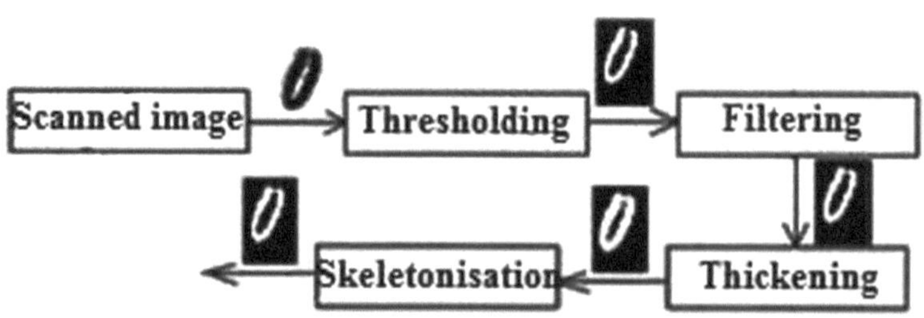

Fig. 3 Block diagram of preprocessing

any character. Figure 3 shows preprocessing sequentially along with the outputs at every step with respect to an MNIST sample of digit ‘zero.’

6 Feature Extraction

Features are aimed at coded representation of a curve. They have been classified broadly into two types based on the parameter of coding, namely distance features and slope features. Both the features have been considered locally and globally.

6.1 Local Distance Features (LDF)

The paper adopts delta distance coding (DDC) to extract local distance features which is basically, a differential approach toward coding a curve based on distance. In differential approach of coding, the difference between magnitudes (pixel intensities in case of images) of two successive samples of a curve is considered that is absolutely lesser than either of the magnitudes. It reduces complexity of the system. While getting difference, more or less equal effects of any local noise on the adjacent magnitudes are canceled out making the noise immunity of the coding system better. Delta distance coding (DDC) has been used to differentiate concavity from convexity while encoding the local curvature in a numeral sample, it reveals the *novelty* of DDC. DDC has been applied both vertically and horizontally.

6.1.1 Vertical Delta Distance Coding (VDDC)

The preprocessed 50 × 50 image is partitioned into four equal sub-images of size 25 × 25 to get local features. Twenty-five samples (a, b, c, d, …) of curves in each of the four sub-image spaces [Fig. 4] are considered, and distance of the samples

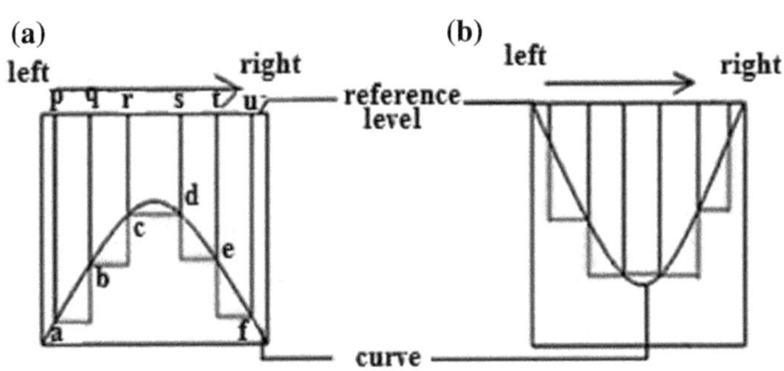

Fig. 4 **a** and **b** Sample points with respective distances

from reference level (pa, qb, rc, sd, …) is noted. The upper edge of each sub-image is conveniently taken as reference line. Delta distance coding for a sample is determined on the basis of comparison between the distance of the current sample and that of the just previous sample.

Let us consider,

d_c = distance of sample 'c' from reference line
d_b = distance of just previous sample 'b' from reference line.

(i) $d_c < d_b$ is encoded as '1'
(ii) $d_c > d_b$ is encoded as '2'
(iii) $d_c = d_b$ is encoded as '0'.

The curves shown in Fig. 4(a) and (b) will be coded as '11022' and '22011', respectively, as per the above logic. The number of feature elements out of VDDC is 25 × 4, i.e., 100

So,

$$F(VDDC) = 100$$

6.1.2 Horizontal Delta Distance Coding (HDDC)

Twenty-five samples are considered from the curve in each sub-images of size 25 × 25. Left edge is conveniently taken as reference line (Fig. 5). The logic followed in HDDC is the same as in VDDC.

The curves in Fig. 5(a) and (b) are coded as '11022' and '22011', respectively. So, HDDC can well differentiate horizontal convexity and concavity. The number of feature elements out of HDDC is 25 × 4, i.e., 100.

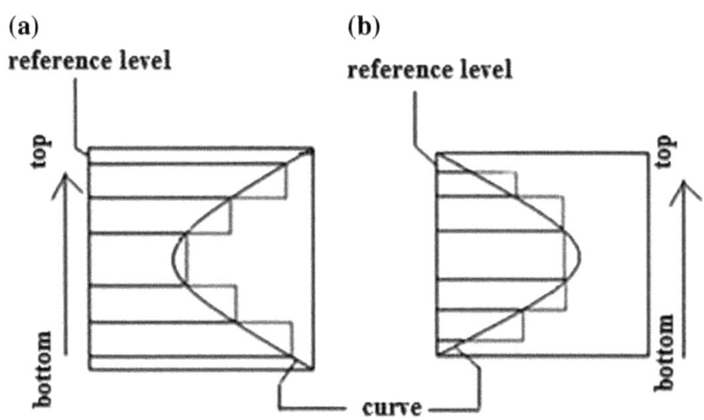

Fig. 5 **a** and **b** Sample points and respective distances from reference level

So

$$F(HDDC) = 100$$

6.2 Global Distance Features (GDF)

In case of oblique or skewed numerals, there remains every possibility of getting more than one samples on each vertical or horizontal line in each sub-image space. It limits the applicability of local distance features to skewed samples. In global distance features, the whole image is considered as one zone and several equal distant samples are collected. The distance of all the samples from the upper edge is summed up together and normalized by dividing the sum by the total number of samples to get a single feature vector from top. In a similar way, feature vectors from bottom, left, and right edges are extracted to get altogether four global distance features. Thus, the feature elements out of global distance feature, i.e., F (GDF) = 4.

6.3 Local Slope Feature (LSF)

The aim of local slope feature is to encode variations in local slope. Figure 6 tells that a concave curve that opens along *X*-axis possesses a gradual increase in local slope as we go along with the curve following the arrow mark. On contrary, a convex curve shows decreasing slope as we go down the curve. Similarly, a concave curve that opens along *Y*-axis shows gradual decrease in slope, whereas a convex curve has increasing slope with reference to *Y*-axis. The current paper finds LSF codes from the feature elements of VDDC and HDDC. Two consecutive 1's in VDDC or HDDC means increasing (positive) local slope and hence coded as positive slope code '3'. On the contrary, two consecutive 2's in VDDC or HDDC means decreasing

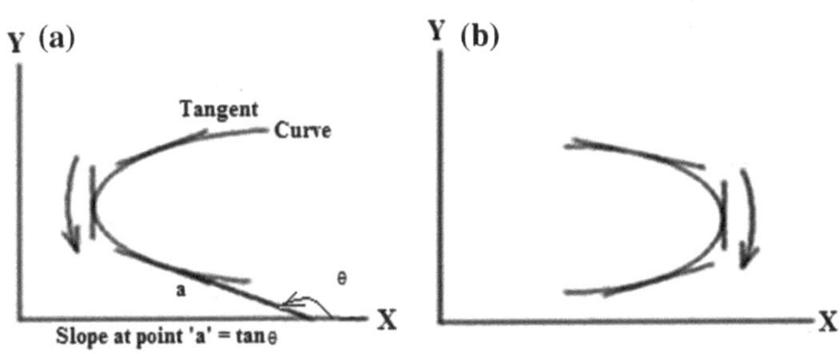

Fig. 6 **a** and **b** Change in slope of curves with axis along *X*-axis

(negative) local slope, and therefore, it is given a negative slope code '4'. So, LSF codes are '3004' and '4003' for LDF codes '11022' and '22011', respectively.

The local slope of a curve is coded as per the scheme shown in Fig. 6. Feature elements out of local slope feature, i.e., F(LSF) = [24 × 4] = 96.

6.4 Global Slope Feature (GSF)

In order to facilitate extraction of slope features from skewed samples, global slope feature has been introduced. Slope, being a vector has both horizontal and vertical components. Horizontal component (S_H) and vertical component (S_V) of slope at a pixel point (x, y) are obtained by using corresponding sobel masks. Magnitude and phase of slope are calculated as per the following relations-

$$\text{Magnitude:}\quad M(x, y) = \sqrt{(S_H^2 + S_V^2)} \tag{1}$$

$$\text{Phase:}\quad \emptyset(x, y) = \tan^{(-1)} \frac{S_V}{S_H} \tag{2}$$

Phase is quantized along eight directions, and total strength in a particular direction is determined by adding up the corresponding magnitudes and then normalized. The results for the angle pairs (0°, ± 180°), (45°, − 135°), (90°, − 90°), and (135°, − 45°) are added up together to get only four feature vectors, i.e., F(GSF) = 4.

The totality of feature elements out of all the four features considered in this paper is as follows:

$$\begin{aligned} O &= [F(LDF)\ F(GDF)\ F(LSF)\ F(GSF)] \\ &= [200\quad 4\quad 96\quad 4] \end{aligned}$$

7 Experiments and Results

Feature extraction methods have been applied on training samples of MNIST dataset, and feature elements are quantized and encoded to eleven symbols to produce observation symbols. The algorithms for training and testing MHMM are Baum–Welch algorithm [27] and Viterbi decoding algorithm, respectively. The HMM model present in Statistics Toolbox of MATLAB version 7.10.0 (R2010a) has been used as the base model. The transition and emission matrices obtained from training constitute the respective model, based on which the classification of testing samples has been executed. Multiple hidden Markov model (MHMM) is trained with observation sequence (O) as per Baum–Welch algorithm. System yields final state transition

Table 2 Phases of experimentation

Phase	Feature(s)	Classifier(s)
1	Distance features	MHMM
2	Slope features	MHMM
3	Distance and slope	MHMM
4	Distance and slope	MHMM-SVM
5	Distance and slope	KNN with hybrid values of 'K'

probability matrix and observation probability matrix for each of ten numeral classes (0–9) and two subclasses of '4' and '7'. Viterbi decoding algorithm is incorporated into decode the sequence of states of the HMM model λ corresponding to observation sequence O obtained from MNIST test samples. The probability of generating the sequence O by the HMM model λ, i.e., P (O/λ) is returned by the system. As per the literature, HMM is a segmentation-free approach and hence, it has an upper hand over other segmentation-based approaches [27–31]. In [32], authors have concluded that with increase in number of states in HMM modeling up to a certain limit, its recognition efficiency increases. The recognition rate declines if the number of states is increased beyond the limit. During experimentation of the paper, the results of proposed MHMM have been observed by increasing the number of states gradually up to thirty-six which produced acceptable accuracy as discussed afterward.

In order to analyze the effectiveness of the features and proposed classifier, the whole experimentation has been presented in the following five phases [Table 2].

7.1 Phase 1, Phase 2, and Phase 3

Multiple hidden Markov model (MHMM) along with distance features only, slope features only, and both features together constitute Phases 1, 2, and 3, respectively, of the experiment. MHMM has been trained with the corresponding feature elements of MNIST training samples, and then classification is done on MNIST test samples to get the phase-wise results as provided in Table 3.

7.2 Phase 4

The proposed system has provision to show the next close output numeral [Table 4] along with the main output. These outputs of each class are further filtered with support vector machine (SVM) in binary mode. SVM, proposed by Vapnik [33, 34], applies structural risk minimization (SRM) principle of statistical learning theory. SVMs have capabilities to generalize well even in high-dimensional spaces with small training samples [35]. They are superior to other traditional approaches

Table 3 Recognition rates of digits in different phases

Digits	Individual percent rate of recognition in		
	Phase 1	Phase 2	Phase 3
0	96.2	96.3	90.6
1	98.6	97.2	95.8
2	96.1	93.7	98.2
3	86.4	90.8	94.4
4	89.5	91.8	89.3
5	90.2	86.8	91.2
6	91.5	81.9	86.9
7	86.5	80.8	97.8
8	88.6	87.3	90.5
9	91.9	86.9	87.5

Table 4 Next closest output numeral(s) against each class

Digits	0	1	2	3	4	5	6	7	8	9
Next close digit (s)	3,7	7	3	8	2,6,7	2,3	2,3,9	3,8	3,7	3,7,8

using empirical risk minimization (ERM) principle like most of neural networks [36]. SVMs erects a hyperplane between the positive (+1) and negative (−1) classes having the largest margin. The current system uses Gaussian RBF kernel function having scaling factor of 1 to pick up one final output. The final rate of recognition of each numeral is calculated considering such final output with proposed cascaded classifier (MHMM-SVM) that have been shown in Table 5. An overall 99.51% rate of recognition has been obtained from the proposed system on MNIST data. The percent increment in recognition accuracy of each digit due to cascaded SVM has

Table 5 Recognition rates of digits with different classifiers

Digits	MHMM	MHMM-SVM cascaded	Increment due to SVM	KNN (K)
0	90.6	99.8	9.2	97.7 (3)
1	95.8	100	4.2	97.7 (1, 3)
2	98.2	100	1.8	85.5 (1, 3)
3	94.4	99.5	5.1	84.6 (1)
4	89.3	98.7	9.4	91.7 (3)
5	91.2	99.1	7.9	90.5 (1, 3)
6	86.9	97.9	11.0	91.6 (3)
7	97.8	99.5	1.7	93.6 (1)
8	90.5	99.8	9.3	78.5 (3)
9	87.5	99.6	12.1	90.7 (1)

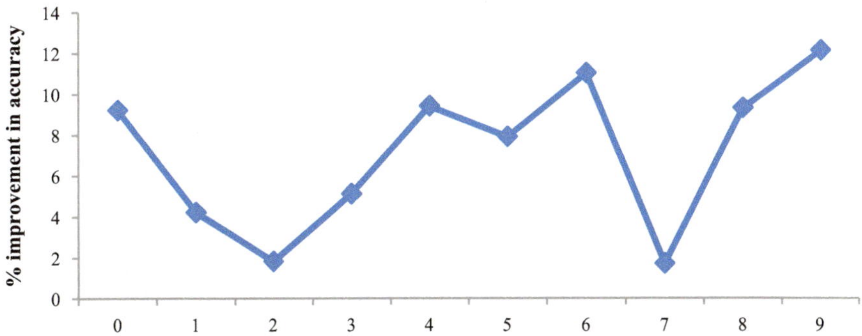

Fig. 7 Percent increase in individual recognition accuracy due to cascaded SVM

been shown in Table 5 and plotted in Fig. 7 to reflect the contribution of support vector machine in the whole process of classification.

7.3 Phase 5

K-nearest neighbor classifier has been trained with both distance features and slope features together, and MNIST test data are classified with K equals 1 and 3. KNN classifier applies Euclidean distance metric to locate the nearest neighbor and majority rule with K-nearest point tie-break to classify a sample point. The higher individual rate out of recognition rates with K equals 1 and 3 is listed in Table 5 with corresponding value(s) of K in brackets along with the outcomes of proposed classifier (MHMM-SVM) cascaded and MHMM alone and respective increments because of SVM. Results in Table 5 and line diagram [Fig. 8] show that (MHMM-SVM) cascaded has outperformed the rest two classifiers.

Reference paper [37] has been considered to compare the results of current paper with those of existing works. The authors have classified Arabic numerals from MNIST database with KNN classifier and derived features based on number of holes, water reservoirs, and fill-hole density in an image. They have reported an average accuracy of 96.94%. The recognition rates of digits in the two systems have been listed in Table 6 and plotted in Fig. 9.

In order to validate the outcomes of the current system with MNIST dataset, the same set of features, i.e., distance features and slope features and same cascaded classification system of MHMM—SVM, have been applied on CENPARMI dataset and the results have been compared to those with MNIST database in Table 7. The overall rate of recognition of the current system on CENPARMI data is 98.8%.

The global rate of recognition the current system is compared in a nutshell chronologically with other works on recognition of English (Arabic) MNIST numeral samples and using different recognition tools (Table 8). Thus, the proposed optical character recognition system contains distance and slope features which have

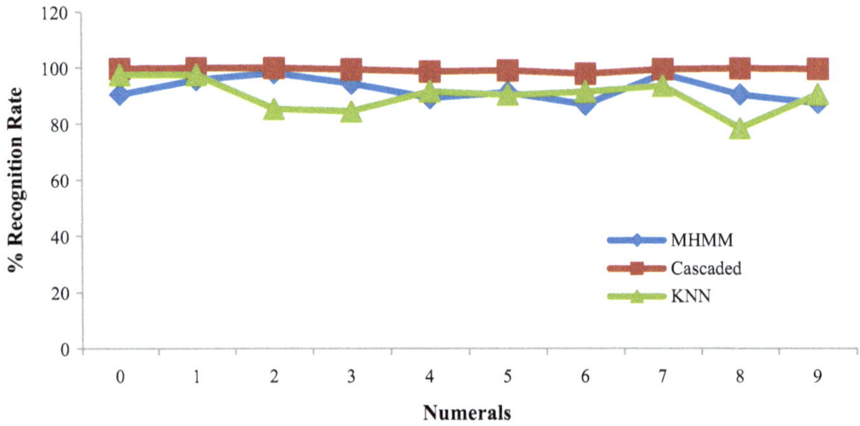

Fig. 8 Line diagrams depicting individual performance of digits with different classifiers

Table 6 Percent accuracies of the two systems at a glance

Digits	Percent recognition accuracy	
	Reference [37]	Proposed paper
0	96.47	99.9
1	97.95	100.0
2	97.26	100.0
3	97.56	99.5
4	97.27	98.9
5	97.47	99.3
6	96.09	98.1
7	95.72	99.8
8	95.71	99.9
9	97.23	99.7

been determined both globally and locally. The methods of extracting these features are devoid of cumbersome calculations, thereby reducing the system complexities and the computation time. The system shows an acceptable accuracy rate of 99.51% on MNIST data which is competitive with the accuracies of the OCR's mentioned in Table 8.

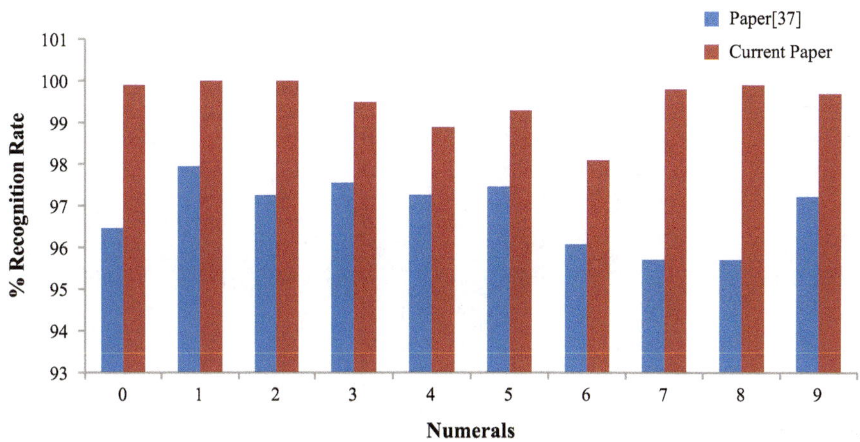

Fig. 9 Columns representing recognition rate of individual digits in both the systems

Table 7 Rates of recognition of numerals in MNIST and CENPARMI dataset

Digits	Outcomes with database	
	MNIST	CENPARMI
0	99.9	98.6
1	100.0	99.5
2	100.0	99.6
3	99.5	98.2
4	98.9	98.2
5	99.3	99.8
6	98.1	98.0
7	99.8	98.3
8	99.9	99.1
9	99.7	98.6

8 Conclusion

The current paper puts up a novel trial to establish a link among several digital domains, i.e., digital communications, digital signal processing, and digital image processing with a purpose to recognize English handwritten numerals. The inspiration behind distance features has been borrowed from delta modulation scheme, a well-known tool in digital communications. SVM complements the extension of well-established classifier hidden Markov model, i.e., multiple hidden Markov model to produce an acceptable percent accuracy of 99.51%. SVM has been applied only to analyze the frequent negative outputs (next close output) for a particular class. For example, the majority of the mis-recognized test Zero (0) samples have produced

Table 8 Comparison with other reported works on MNIST data

1st author	Year	Ref.	Classifier	% rate
Bhattacharya	2009	[4]	MLP	98.79
A. Ko	2009	[41]	HMM–LOOT	98.88
N. Yu	2012	[38]	SVM	99.10
Z. Zhao	2013	[39]	Neural network	91.24
Karungaru	2013	[2]	Neural network	97.10
H. Cecotti	2013	[40]	Neural network	93.01
U. Babu	2014	[37]	KNN	96.94
Qacimy	2014	[3]	SVM	98.76
S. Celar	2015	[10]	MLP	94.76
R. Janrao	2016	[42]	KNN	93.65
Proposed	2019	NA	HMM-SVM cascaded	99.51

outcomes in favor of '3' and '7' (Table 4). Only such samples are classified once again by means of SVM, thereby reducing the classification burden. Thus, SVM has been used as a post-processor and not as the main classifier. In Table 5, MHMM—SVM cascaded shows more or less uniform recognition rates for all the digits except for '6'. The close resemblance with (2, 3, and 9) still hampers its result. The poor results of some digits are also because of their badly written samples in CENPARMI and MNIST datasets. Some more promising features and efficient post-processing will be used in the follow-up works to elevate the individual digit-wise as well as the global rate of recognition of the whole system.

References

1. Liu CL, Nakashima K, Sako H, Fujisawa H (2004) Handwritten digit recognition: Investigation of normalization and feature extraction techniques. Pattern Recogn 37(2):265–279
2. Karungaru S, Terada K, Fukumi M (2013) Hand written character recognition using star-layered histogram features. In: SICE annual conference, pp 1151–1155, Nagoya University, Japan
3. Qacimy B, Kerroum M, Hammouch A (2014) Feature extraction based on DCT for handwritten digit recognition. Int J Comput Sci 11(2): 27–33
4. Bhattacharya U, Chaudhuri BB (2009) Handwritten numeral databases of Indian scripts and multistage recognition of mixed numerals. IEEE Trans Pattern Anal Mach Intell 31(3): 444–457
5. Dhande PS, Kharat R (2018) Character recognition for cursive English handwriting to recognize medicine name from doctor's prescription. In: International conference on computing, commnication, conntrol and automation (ICCUBEA), Pune, India, IEEE
6. Chakraborty B, Shaw B, Aich J, Bhattacharya U, Parui S (2018) Does deeper network lead to better accuracy: A case study on handwritten Devanagari characters. In: 13th IAPR international workshop on document analysis system (DAS), Vienna, Austria, IEEE
7. Choudhury A, Rana HS, Bhowmik T (2018) Handwritten Bengali numeral recognition using HoG based feature extraction algorithm. In: 5th international conference on signal processing and integrated networks (SPIN), Noida, India, IEEE

8. Mathur A, Pathare A, Sharma P, Oak S (2019) AI based reading system for blind using OCR. In: International conference on electronics, communication and aerospace technology (ICECA), Coimbatore, India, IEEE
9. Chen C, Shih P, Liang W (2016) Integration of orthogonal feature detectors in parameter learning of artificial neural networks to improve robustness and the evaluation on handwritten digit recognition tasks. In: IEEE international conference on acoustics, speech and signal processing (ICASSP)
10. Celar S, Stojkic Z, Seremet Z (2015) Classification of test documents based on handwritten student id's characteristics. In: 25th DAAAM international symposium on intelligent manufacturing and automation, DAAAM, procedia engineering (Elsevier), pp 782–790
11. LeCun Y, Bottou L, Bengio Y (1998) Gradient-based learning applied to document recognition. Proc IEEE 86(11): 2278–2324
12. Chen G, Li Y, Srihari S (2016) Word recognition with deep conditional random fields. In: International conference on image processing (ICIP), IEEE
13. Xiao X, Suen C (2012) A novel hybrid CNN-SVM classifier for recognizing handwritten digits. Pattern Recognit 45(4): 1318–1325
14. Lopes G, Silva D, Rodrigues A, Filho P (2016) Recognition of handwritten digits using the signature features and optimum path forest classifier. IEEE Lat Am Trans 14:2455–2460
15. Agarwal A, Goswami S (2016) An efficient algorithm for automatic car plate detection and recognition. In: Second international conference on computational intelligence and communication technology (CICT), IEEE Explore, pp 644–648
16. Zhang P, Bui T, Suen C (2005) Hybrid feature extraction and feature selection for improving recognition accuracy of handwritten numerals. In: Eighth international conference on document analysis and recognition (ICDAR'05), Computer Society IEEE
17. Wu X, Tang Y, Sabourin R, Bu W (2014) Offline text independent writer identification based on scale invariant feature transform. IEEE Trans Inf Forensics Secur 9(3): 526–536
18. Wang B, Zhang L, Wang X (2013) A classification algorithm in li-K nearest neighbor. In: Fourth global congress on intelligent systems, IEEE
19. Gil A, Filho C, Costa M (2014) Handwritten digit recognition using SVM binary classifiers and unbalanced decision trees. In: Image analysis and recognition (springer), pp 246–255
20. Elleuch M, Lahiani H, Kherallah M (2015) Recognizing Arabic handwritten script using support vector machine classifier. In: 15th international conference on intelligent systems design and applications (ISDA)
21. Boukharoubal A, Bennia A Novel feature extraction technique for the recognition of handwritten digits. In: Applied computing and informatics, Saudi computer society
22. Osadchy M, Keren D, Raviv D (2016) Recognition using hybrid classifiers. IEEE Trans Pattern Anal Mach Intell 38(4): 759–771
23. Henge SK, Rama B (2017) OCR-mirror image reflection approach: Document back side character recognition by using neural fuzzy hybrid system. In 7th international advance computing conference (IACC), Hyderabad, India, IEEE, July 2017
24. Aziz T, Rubel AS, Salekin S, Kushol R (2018) Bangla handwritten numeral character recognition using directional pattern. In 20th international conference of computer and information technology (ICCIT), Dhaka, Bangladesh, IEEE, Feb 2018
25. Ashiquzzaman A, Tushar AK (2017) Handwritten Arabic numeral recognition using deep learning neural networks. In: International conference on imaging, vision & pattern recognition (icIVPR), Dhaka, Bangladesh, IEEE, Apr 2017
26. Taktak M, Triki S, Kamoun A (2018) 3D handwriting characters recognition with symbolic-based similarity measure of gyroscope signals embedded in smart phone. In: 14th international conference on computer systems and applications (AICCSA), IEEE, Mar 2018
27. Rabiner LR (1998) A tutorial on hidden Markov models and selected applications in speech recognition. Proc IEEE 77(2): 257–286
28. Abed HEl, Margner V (2011) ICDAR—Arabic handwriting recognition competition. Int J Doc Anal Recognit 14(1): 3–13

29. El-Hajj R, Likforman-Sulem L, Mokbel C (2005) Arabic hand-writing recognition using baseline dependent features and hidden Markov modeling. In: Eighth international conference on document Analysis and recognition, pp 893–897
30. Mokbel C, Abi Akl H, Greige H (2002) Automatic speech recognition of Arabic digits over telephone network. In: Proceedings of research trends in science and technology, RSTS'02
31. Natarajan P, Saleem S, Prasad R (2008) Multilingual off-line handwriting recognition using hidden Markov models: A script independent Approach. Springer book chapter on Arabic and Chinese handwriting recognition, pp 235–250
32. Abou-Moustafa KT, Cheriet M, Suen CY (2004) On the structure of hidden Markov models. Pattern Recognit Lett, Elsevier 25: 923–931
33. Cortes C, Vapnik V (1995) Support vector networks. Mach Learn 20: 273–297
34. Vapnik V (1995) The nature of statistical learning theory. Springer, New York
35. Jonsson K, Kittler J, Matas YP (2002) Support vector machines for face authentication. J Image Vis Comput 20: 369–375
36. Lu J, Plataniotis KN, Ventesanopoulos AN (2001) Face recognition using feature optimization and v-support vector machine. In: IEEE neural networks for signal processing, pp 373–382
37. Babu U, Venkateswarlu Y, Chintha A (2014) Handwritten digit recognition using k-nearest neighbour classifier. In: Proceedings of world congress on computing and communication technologies, IEEE
38. Yu N, Jiao P (2012) Handwritten digits recognition approach research based on distance & Kernel PCA. In: IEEE fifth international conference on advanced computational intelligence (ICACI), pp 689–693
39. Zhao Z, Liu CL, Zhao M (2013) Handwriting representation and recognition through a sparse projection and low-rank recovery framework. In: International joint conference on neural networks (IJCNN), pp 1–8
40. Cecotti H, Vajda S (2013) A radial neural convolutional layer for multioriented character recognition. In: 12th ICDAR, IEEE
41. Ko A, Cavalin P, Sabourin A (2009) Leave-one-out-training and leave-one-out-testing hidden markov models for a handwritten numeral recognizer: The implications of a single classifier and multiple classifications. IEEE Trans Pattern Anal Mach Intell 31(12)
42. Janrao R, Dighe, D (2016) Handwritten English character recognition using LVQ and KNN. In: International journal of engineering sciences and research technology, pp 904–912

Chapter 14
Clustering and Employability Profiling of Engineering Students

Monika Gupta Vashisht and Reena Grover

1 Introduction

Non-employability of eighty percent of engineers is an alarming situation for the nation [1]. The situation is further worsened with the information that there occurred 'no change' during the last nine years. The Indian education system revolves more around university-prescribed curriculum, and emphasis is laid more on theoretical aspects. Students' performance prediction rests on educational data mining (EDM). The engineering institutes are still struggling to inculcate appropriate skills among engineering students to enhance employability [9].

1.1 Objective of the Study

The objective of the study undertaken is:

To profile employable engineering students on the basis of demographic variables.

2 Review of Literature

An attempt has been made to enhance employability skills among engineering graduates utilizing appropriate methodologies while teaching [8]. We need to focus on enhancing employability though we have been successful in enhancing performance academically [6].

M. Gupta Vashisht (✉)
Chandigarh Business School of Administration, Chandigarh Group of Colleges, Mohali, Punjab, India

R. Grover
SRM Institute of Science and Technology, Ghaziabad, Uttar Pradesh, India

S. Jain et al. (eds.), *Advances in Computational Intelligence Techniques*, Algorithms for Intelligent Systems,
https://doi.org/10.1007/978-981-15-2620-6_14

Karli [4] observed significant differences between perceptions of employability and education level, but not any such difference regarding gender and communicating in a foreign language. Chithra [2] made an attempt to know the employers' and employees' perceptions regarding employability skills and found a significant difference between these. The students' unemployability might be an outcome of such disparity. Ramisetty and Desai [7] found the gap between competence at entry-level employability skills and competencies required.

3 Research Methodology

The study is focused on profiling of employable engineering students based on their demographics. For this, the students who have opted for CSE, IT, ECE and ME courses and falling in different age groups have been asked to mention their level of agreement toward employability skills via an open-ended questionnaire.

Measures: The current study undertaken is focused on profiling employability skills motives of engineering students on the basis of demographic variables such as gender and age group. The instrument was designed using "Employability Appraisal Scale (EAS)" developed by Llinares-Insa et al. [5] (Appendix).

The survey included the following sections:

- Questions related to student demographics—gender, age group and course of study;
- 35 items to measure the employability skills of engineering students.

The data was collected using five-point Likert scale of agreement with '5' reflecting 'strong agreement' to '1' reflecting 'strong disagreement.'

Sampling Design: The universe of the study comprised all engineering students (above 18 years of age) residing in India. The survey (target) population included all engineering students in Mohali (Punjab) and Ghaziabad (UP) in India.

The non-probability judgment and convenience sampling were used. Fifteen respondents comprised the final sample.

- **Description of Sample**: The management students supported the researchers in gathering relevant data.
- **Demographics**: Gender (course-wise) and age group (course-wise) (Table 1).

The combined sample consisted of more than half (54.4%) of the respondents pursuing CSE, another one-fourth (28%) pursuing ECE, whereas the other 12% were pursuing IT and remaining 5.6% pursuing ME course in engineering program.

In summary, the typical sample respondent profile can be described as follows.

The engineering students were mainly male in the age group of 20–21 years and enrolled in CSE or ECE courses.

Table 1 Sample frequencies of engineering students

Descriptive	Frequency	Percentage
Course of study		
CSE	68	54.4
IT	15	12
ECE	35	28
ME	7	5.6
Total	125	
Age group (in years)		
CSE (18–19)	8	6.4
CSE (19–20)	12	9.6
CSE (20–21)	30	24
CSE (21–22)	18	14.4
IT (18–19)	1	0.8
IT (19–20)	2	1.6
IT (20–21)	8	6.4
IT (21–22)	4	3.2
ECE (18–19)	2	1.6
ECE (19–20)	3	2.4
ECE (20–21)	18	14.4
ECE (21–22)	12	9.6
ME (18–19)	1	0.8
ME (19–20)	1	0.8
ME (20–21)	3	2.4
ME (21–22)	2	1.6
Total	125	
Gender		
CSE Male	39	31.2
CSE Female	29	23.2
IT Male	9	7.2
IT Female	6	4.8
ECE Male	22	17.6
ECE Female	13	10.4
ME Male	6	4.8
ME Female	1	0.8
Total	125	

4 Statistical Analysis

The study tested statistically the engineering students' perceptions toward employability skills based on gender and age group. The statistical significance was set at a level of 0.05. Cluster analysis via Ward's method was used.

4.1 Results and Findings of the Study

Cluster analysis has been used via SPSS Statistics, 16.0.1 version for profiling the engineering students. A new student typology, named as Practicable Students and Impracticable Students, has been evolved. The study revealed a statistically significant relationship between the *Employability Skills Motives*-based clusters (Practicable Students and Impracticable Students) based on select demographic variables: gender and age group under study.

For cross-tabulating the data, a chi-squared test has been administered in order to profile the two segments. The statistically significant differences were found (Refer to Tables 2, 3 and 4).

4.1.1 Summary of Results of Chi-Squared Tests of Independence on Engineering Student's Clusters

In summary, we can say that there is a statistically significant relationship between the *Employability Skills Motives*-based clusters (Practicable Students and Impracticable Students) and the select demographic variables: gender and age group under study.

Table 2 Frequencies of engineering students

Descriptive	Practicable students (cluster II)	Impracticable students (cluster I)	Total sample frequency
Course			
CSE	53	15	68
IT	12	3	15
ECE	27	8	35
ME	5	2	7
Total	97	28	125

Table 3 Gender * engineering students' clusters

Gender * cluster crosstabulation

			Cluster		Total
			1	2	
Gender	1	Count	17	59	76
		Expected count	17.0	59.0	76.0
		% within gender	22.4%	77.6%	100.0%
		% within cluster	60.7%	60.8%	60.8%
		% of total	13.6%	47.2%	60.8%
	2	Count	11	38	49
		Expected count	11.0	38.0	49.0
		% within gender	22.4%	77.6%	100.0%
		% within cluster	39.3%	39.2%	39.2%
		% of total	8.8%	30.4%	39.2%
Total		Count	28	97	125
		Expected count	28.0	97.0	125.0
		% within gender	22.4%	77.6%	100.0%
		% within cluster	100.0%	100.0%	100.0%
		% of total	22.4%	77.6%	100.0%

Age group: 1—male, 2—female
Cluster 1: practicable students, 2: impracticable students

5 Discussions

The study has been concluded with the objective of profile engineering students on the basis of demographic variables such as gender and age group. Clustering analysis revealed two types of clusters of such students. One group is labeled as 'Practicable Students.' It comprises all such students from various courses who are focused on learning employability skills. Students pursuing CSE and ECE courses in engineering programs are more focused on learning employability skills (i.e., Practicable Students) than those pursuing IT and ME courses.

The other group comprises 'Impracticable Students' who are not focused on learning employability skills (Table 5).

Engineering Institute Strategies for Enhancing Employability Skills Among Students: The institutions and universities offering engineering programs in various courses need to focus on technical knowledge, soft skills as well as inculcating right attitude among students since beginning, i.e., during the first year itself when the students enroll themselves in the institute or universities. They need to appoint professionals for the purpose of who can groom them beyond prescribed university curricula. The regular industrial visits, guest lectures, alumni interactions, group discussions, mock interviews, case studies, role-plays and similar activities need to

Table 4 Age group * engineering students' clusters

Age_Group * cluster crosstabulation

			Cluster		Total
			1	2	
Age_Group	1	Count	15	53	68
		Expected count	15.2	52.8	68.0
		% within Age_Group	22.1%	77.9%	100.0%
		% within cluster	53.6%	54.6%	54.4%
		% of total	12.0%	42.4%	54.4%
	2	Count	3	12	15
		Expected count	3.4	11.6	15.0
		% within Age_Group	20.0%	80.0%	100.0%
		% within cluster	10.7%	12.4%	12.0%
		% of total	2.4%	9.6%	12.0%
	3	Count	8	27	35
		Expected count	7.8	27.2	35.0
		% within Age_Group	22.9%	77.1%	100.0%
		% within cluster	28.6%	27.8%	28.0%
		% of total	6.4%	21.6%	28.0%
	4	Count	2	5	7
		Expected count	1.6	5.4	7.0
		% within Age_Group	28.6%	71.4%	100.0%
		% within cluster	7.1%	5.2%	5.6%
		% of total	1.6%	4.0%	5.6%
Total		Count	28	97	125
		Expected count	28.0	97.0	125.0
		% within Age_Group	22.4%	77.6%	100.0%
		% within cluster	100.0%	100.0%	100.0%
		% of total	22.4%	77.6%	100.0%

Course of study: 1—CSE, 2—IT, 3—ECE, 4—ME
Cluster 1: practicable students, 2: impracticable students

Table 5 Segmentation profiles of engineering students' clusters

Characteristics	Practicable Students	Impracticable Students
Employability skills' motives	Focused on learning employability skills	Not focused on learning employability skills
Profession	42.4% CSE	12.0% CSE
	9.6% IT	2.4% IT
	21.6% ECE	6.4% ECE
	4% ME	1.6% ME

be incorporated in reality. The area where institutes and universities are lacking is paying attention to every individual besides sharing feedback and scope and ways of improvement on a regular basis. The most important thing is 'knowing your student,' i.e., which skill set they already possess, what they want to become, what efforts they have been putting up on regular basis to achieve their goal of becoming employable. The study revealed that there is a niche of students who are impracticable. The reasons beneath need to be explored further, and accordingly, employability skill enhancement strategies may be propounded.

6 Conclusion

The study has been concluded with profiling of engineering students from various courses as 'Practicable Students' and 'Impracticable Students' on the basis of gender and age group as demographic variables. Students are unable to perform well during interview and to fetch appropriate jobs due to insufficient direction and counseling [1]. It is recommended in the report that entire focus should be on improving education quality for which new programs and FDPs are required. The stakeholders may be duly incentivized for promoting employability. Also, academia–industry and academia interface needs to be strengthened.

7 Limitations of the Study

- The study covered only the respondents pursuing engineering in selected courses, other engineering students may exhibit different behavior.
- The exploratory study was conducted in Mohali, Punjab, and Ghaziabad, UP, India; the engineering students' perceptions and attitudes might vary in other regions.

8 Further Scope of Study

The study could be carried out in other regions throughout the world with larger sample size and in other courses and programs. The current study provides the basis for the same.

Appendix

Employability Appraisal Scale (EAS)

1	I achieve what I set out to do
2	I have confidence in my own opinions, even if they are different from other people's
3	I get to work when I decide what I want to do
4	I get bored with doing daily activities
5	I have a tendency to change activities during the day without a specific reason
6	I have a bad appearance and I think that is why I can't find a job
7	I do not persevere when I have to perform a long and difficult task
8	When I need to know something at work I usually ask or ask to be taught
9	I can design a good plan of action when I have to do something important related to my studies or my work
10	My training is insufficient to work as a professional
11	I do not have enough experience to be hired
12	I get involved in what I do, and I am enthusiastic about the tasks I undertake
13	For me, it is more important to feel good about myself than to receive the approval of others
14	There are other professionals better prepared to work than I am
15	I consider myself effective in my work
16	I have a tendency to leave things until the last minute
17	I find it difficult to control my anger
18	When I have to do an activity I take a long time to get going
19	I have problems organizing the things I have to do
20	I am responsible for my actions and decisions
21	Some things annoy me a lot
22	I like to learn new things about my work even if it's about small details
23	I have the impression that I cannot do the activities that need to be done every day
24	I can't find a job because I don't know how to look for one
25	I am a practical person. I know what I have to do and I do it
26	I can't find a job because I lack the ability to express myself and relate to other people
27	I can organize my time to make the most of it
28	I can't find a job because I lack self-confidence
29	I am persistent and tenacious. I finish what I start
30	I can't find a job because I have to be more persistent when I search for employment and not get discouraged
31	I can't find a job because I don't keep up with my profession and I'm not competent
32	I get angry easily
33	I have a bad temper

(continued)

(continued)

34	I consider myself a person with initiative for beginning tasks, making decisions, or solving problems
35	I view changes as an opportunity to learn, and not as a difficulty

References

1. Aspiring Minds' National Employability Report-Engineers (2019)
2. Chithra R (2013) Employability skills—a study on the perception of the engineering students and their prospective employers. Glob J Manag Bus Stud 3(5):525–534
3. https://www.counterview.net/2019/03/skill-india-80-of-engineers-not.html
4. Karli U (2016) Adaptation and validation of self-perceived employability scale: an analysis of sports department students and graduates. Educ Res Rev 11(8):848–859
5. Llinares-Insa LI, González-Navarro P, Zacarés-González JJ, Córdoba-Iñesta AI (2018) Employability Appraisal Scale (EAS): development and validation in a Spanish sample. Front Psychol 9(article 1437):1–11
6. Mishra T, Kumar D, Gupta S (2017) Students' performance and employability prediction through data mining: a survey. Indian J Sci Technol 10(24):1–6
7. Ramisetty J, Desai K (2017) Measurement of employability skills and job readiness perception of post-graduate management students: results from a pilot study. Int J Manag Soc Sci 5(8):82–94
8. Selvi R, Anitha P, Padmini P (2018) A study on employability of engineering graduates using data mining techniques. Int J Eng Sci Inven (IJESI): 12–16
9. Thakar P, Mehta A, Manisha S (2016) Cluster model for parsimonious selection of variables and enhancing students' employability prediction. Int J Comput Sci Inf Secur (IJCSIS)14(12):611–618

Chapter 15
Low-Cost Fractal MIMO Antenna Design for ISM Band Using ANN

Balwinder S. Dhaliwal, Gurpreet Kaur, Simranjit Kaur and Suman Pattnaik

1 Introduction

Multiple-inputs multiple-outputs (MIMO) is the advanced technology for achieving higher throughputs, higher data rates and large bandwidth characteristics in the wireless communication development. The MIMO was first projected by the pioneer Foschini, to improve the data rates by employing MIMO at the transmitter as well as receiver side of a communication system [1]. In wireless communications systems, MIMO technique has been used for sending and receiving more than one signal at the same time [2]. MIMO wireless communication has many applications including Wi-Fi, massive MIMO and long-term evolution [3]. The use of multiple antennas in MIMO systems increases the data rate and power levels [4], however, the design of antennas for MIMO systems is a challenging and several new geometries have been proposed in recent years. Nigam and Kumar [5] designed and analyzed a MIMO system using microstrip line fed compact antennas for ISM band frequencies. Anuvind and Joseph designed and simulated a 2×2 MIMO microstrip patch antenna array for 2.45 GHz applications [6]. Antenna elements having semi-printed structure and fractal shapes operating in 2.45 GHz ISM band and in 5–6 GHz band have been presented in [7].

The ANN models have been used by many researchers to estimate the antenna dimensions for the desired frequencies because ANN models are easy to apply and provide accurate results. Singh [8] introduced a method to compute the resonant frequency of rectangular patch antenna by resilient back-propagation, feed-forward

B. S. Dhaliwal (✉)
National Institute of Technical Teachers Training and Research, Chandigarh 160019, India

G. Kaur · S. Kaur
Guru Nanak Dev Engineering College, Ludhiana, India

S. Pattnaik
Sri Sukhmani Institute of Engineering & Technology, Dera Bassi, Punjap, India

S. Jain et al. (eds.), *Advances in Computational Intelligence Techniques*,
Algorithms for Intelligent Systems,
https://doi.org/10.1007/978-981-15-2620-6_15

back-propagation algorithm, radial basis functions and Levenberg-Marquardt of ANN. Arora and Dhaliwal [9] proposed ANN for dual-band frequency estimation of coaxial fed fractal patch antenna. Authors of [10] designed a pre-fractal antenna using BFO and ensemble of ANN for ISM band applications [10]. In [11], the design of a fractal antenna based on the circular base shape has been presented for 5.8 GHz band applications using PSO-ANN algorithm. Results of pentagonal fractal antenna have been observed by ANN in [12].

In this proposed work, a low-cost two-element MIMO fractal antenna has been designed for ISM band applications. The antenna design is explained in the next section. An ANN synthesis model has been used for designing antenna which is also briefly summarized in Sect. 2. In Sect. 3, the simulated and measured results are described. The manuscript has been concluded in Sect. 4.

2 Proposed Antenna Design and ANN Model

The fractal antenna geometry discussed in this presented work has been taken from [10] and [13] but with the change in substrate material. RT-Duroid substrate material of substrate thickness (h) equal to 3.175 mm and dielectric constant ($\in_r$) equal to 2.2 has been used in reference antennas of [10] and [13]. In presented work, FR4 substrate material with $h = 1.47$ mm, $\in_r = 4.3$ and $\tan\delta = 0.017$ has been used to develop low-cost antenna.

The proposed antenna has been developed by starting with rectangular antenna as base geometry. Then, an elliptical slot is created in base rectangle shape and subsequently a rectangular shape is inserted in the elliptical slot such that corners of inserted rectangular shape have contact with the borderline of the elliptical cut. The identical process has been applied to obtain second iteration of the fractal geometry. This shape has size miniaturization characteristics [10]. The opposite corners of this design are truncated for bandwidth enhancement. Also, the straight boundaries of the base rectangle are replaced by Koch curve to achieve more miniaturization of antenna [13]. The final geometry of the proposed hybrid fractal antenna has been presented in Fig. 1.

The antenna geometry of the [10] has been proposed for 2.45 GHz applications. However, the addition of Koch curve boundaries and change of substrate material causes the shifting of operating frequency from 2.45 GHz. To estimate the optimal dimension for the desired operation at 2.45 GHz frequency, an ANN synthesis model has been developed. The ANN model estimates the outermost dimensions, i.e., length (L) and width (W) of presented rectangular geometry for the desired resonant frequency (f_r) applied as input as shown in Fig. 2. The inner dimensions are calculated with the guidelines provided in [10].

ANN requires a data set of sufficient size for development of the model of the proposed. Therefore, 36 geometries of proposed antenna with various dimensions of L and W have been simulated and respective f_r values have been recorded. The data values for the presented antenna have been estimated by IE3D simulator. For ANN

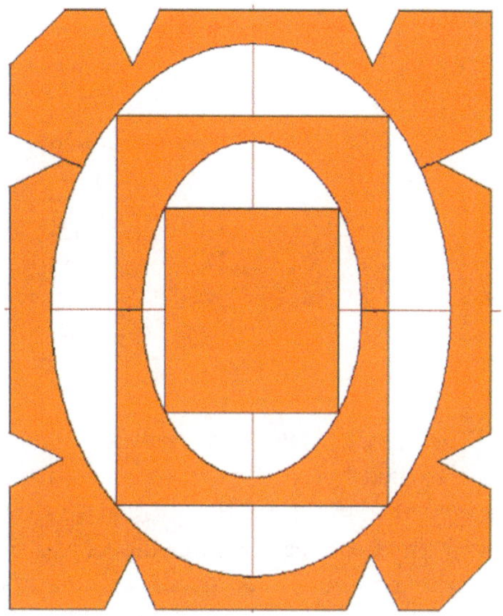

Fig. 1 Design of hybrid fractal antenna

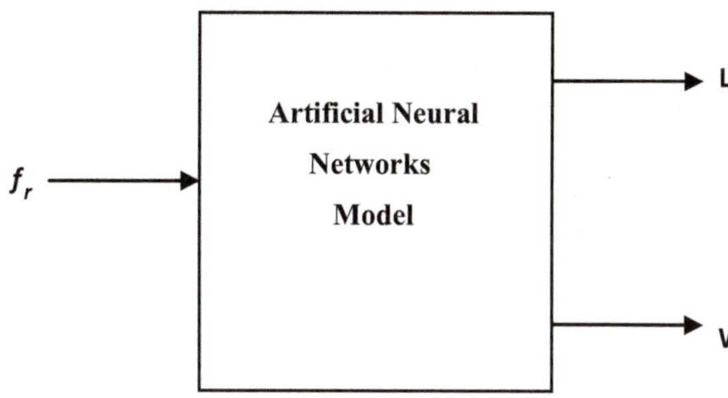

Fig. 2 Block diagram of ANN model for presented fractal geometry

synthesis model development, this data set has been used with f_r taken as input, whereas L and W used as outputs as shown in Fig. 2.

3 Results and Discussion

The 30 pairs of input and output values from data set have been employed for the training of proposed ANN model. The model has been designed 15 neurons in the one hidden layer employed for the design. The trained model has been tested using test data input values and the predicted outputs for these inputs are compared with the expected values as shown in Table 1 which depicts that ANN model has predicted dimensions for desire resonant frequency with reasonable accuracy.

Table 1 Test results of ANN synthesis model

S. no	Test input	Target O/P		ANN O/P		Errors in	
	f_r	L	W	L	W	L	W
1	2.08	28.00	34.20	24.44	32.94	0.55	1.25
2	2.17	24.65	30.07	24.57	30.46	0.07	0.39
3	2.28	23.50	28.67	23.99	29.07	0.49	0.40
4	2.61	21.95	26.80	20.35	24.99	1.59	1.80
5	2.66	20.40	25.10	20.17	24.82	0.22	0.27
6	2.92	18.00	22.90	19.01	23.29	1.01	0.39

The trained ANN model has been then used to estimate the values of L and W of presented hybrid fractal geometry for ISM band applications with center frequency of 2.45 GHz. So, 2.45 GHz has been applied at the input of ANN model and the model generated the outputs as $L = 22.01$ mm and $W = 26.83$ mm for this input. Therefore, the proposed antenna of these dimensions is expected to radiate at 2.45 GHz frequency. This designed antenna geometry has been then employed to implement a two-element MIMO antenna for ISM band applications as shown in Fig. 3. To develop the MIMO antenna design with small mutual coupling between the elements, 10 mm distance has been used between two-element antennas that are placed in parallel on FR4 material.

The ANN results have been validated using simulation results and experimental measurements. Two antenna elements of ANN predicted dimensions have been simulated to validate the precision of ANN model outputs. An inter-element distance of 10 mm has been taken for this simulation. The S-parameter results are shown in Fig. 4 for this optimized MIMO fractal antenna and these plots show a perfect matching of S_{11} and S_{22} graphs and also identical shape of S_{12} and S_{21} graphs.

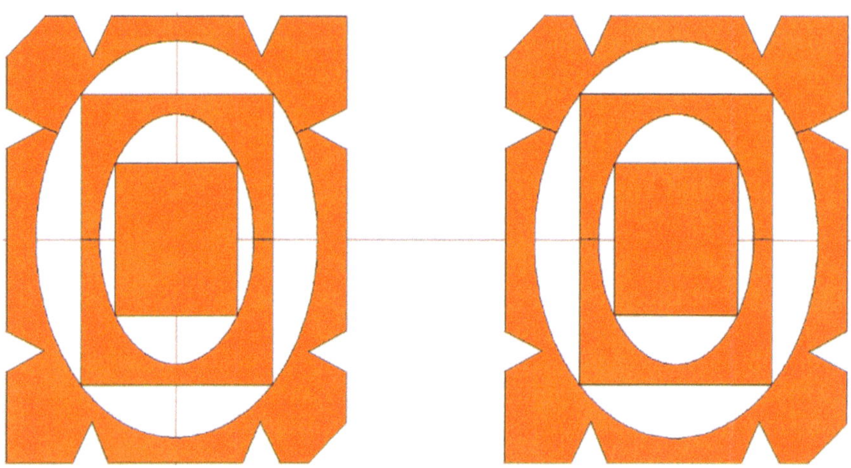

Fig. 3 Proposed antenna design optimized for 2.45 GHz

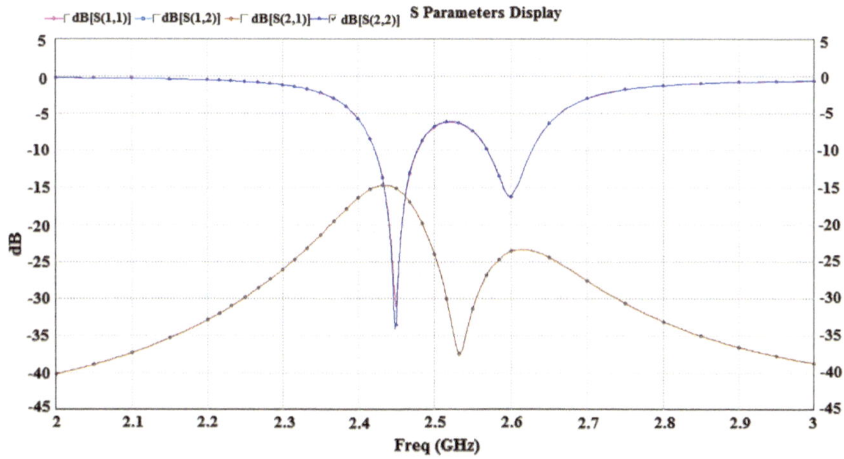

Fig. 4 S-parameter plots of designed MIMO antenna for ISM band

S-parameter graphs of Fig. 4 depict that the fundamental resonant frequency of the designed antenna geometry is 2.4503 GHz and it implies that dimensions provided by the ANN model are accurate and the designed antenna can be used for 2.45 GHz ISM band applications. However, the designed antenna also operates efficiently at 2.60 GHz; therefore, the presented antenna geometry is a two-band MIMO antenna. However, the major objective of the current work is to design MIMO antenna for ISM band frequencies.

Fig. 5 Snapshot of prototype antenna

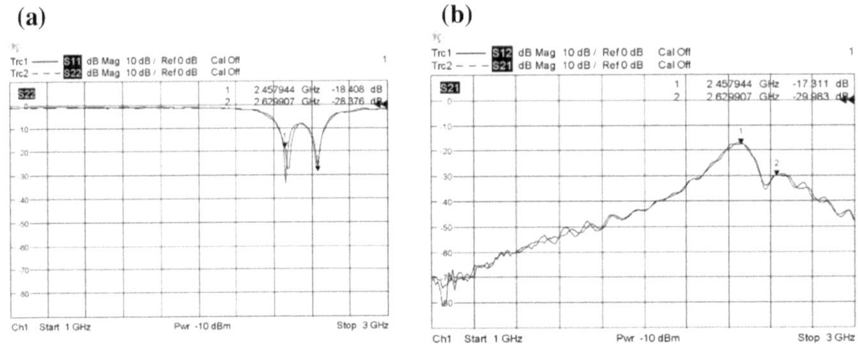

Fig. 6 Measured results of fabricated prototype (a) S_{11} and S_{22} plots (b) S_{12} and S_{21} plots

Table 2 Comparison of simulated and measured results

Characteristics parameters	Simulated results	Measured results
f_r (GHz)	2.4503	2.457
S_{11} (dB)	−30.86	−28.44
S_{12} (dB)	−15.19	−17.39
S_{21} (dB)	−15.19	−17.69
S_{22} (dB)	−30.86	−33.61

After the validation of antenna performance by simulator, the prototype of the presented antenna has been fabricated and experimental results have been measured. Figure 5 shows the top view of fabricated prototype. The measurement of S-parameters has been done using vector network analyzer of R & S make.

The measured S-parameter results have been given in Fig. 6 which depict that the simulated and experimental results are in good agreement. The fundamental resonant frequency of fabricated antenna has been measured as 2.457 GHz and the second resonant frequency has been measured as 2.629 GHz. The comparison of the simulated and measured results for ISM band has been shown in Table 2 which shows a good agreement of all parameters. The main reason for the slight mismatch of magnitudes of S-parameters is imperfect coaxial connectors.

The gain plot of the designed antenna is shown in Fig. 7 which depict that antenna has a gain 2 dBi at both resonant frequencies.

4 Conclusion

RT-Duroid material of substrate of the reference antenna has been changed to FR4 material to design a low-cost miniaturized antenna geometry. An ANN synthesis

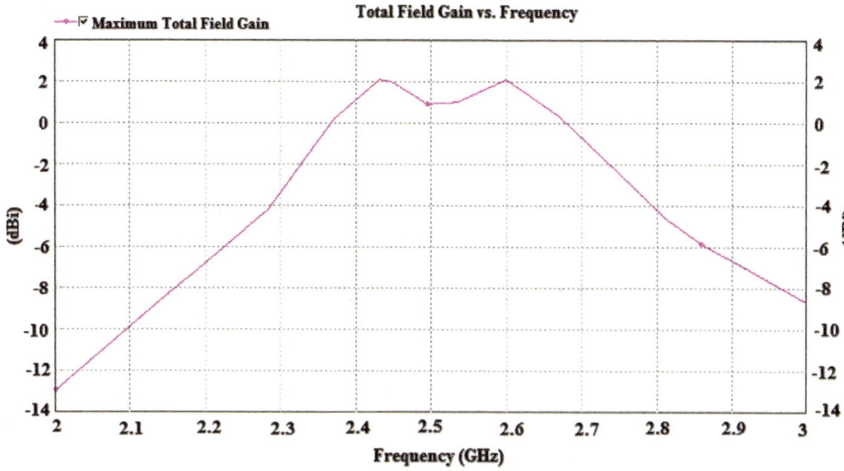

Fig. 7 Gain plot of designed antenna

model has been developed to design the selected geometry for 2.45 GHz applications. The trained ANN model estimated the values of antenna dimensions which have been validated using software and measurements. Simulated and measurement results of the designed MIMO antenna geometry have good matching which verifies the accuracy of the adopted ANN approach. The 2.45 GHz fundamental resonant frequency of proposed geometry is suitable for ISM band. The second resonant frequency of 2.60 GHz makes the presented antenna a multiband design.

References

1. Krishna KSR, Babu KJ, Narayana JL, Reddy LP, Subrahmanyam GV (2012) Artificial neural network approach for analyzing mutual coupling in a rectangular MIMO antenna. Front Electr Electron Eng 7(3):293–298
2. Kaur K, Dhaliwal BS (2017) Analysis of MIMO antenna using artificial neural network. In: International Conference on Soft Computing Applications in Wireless Communication, pp 184–186
3. Soltani S, Murch RD (2015) A compact planar printed MIMO antenna design. IEEE Trans Antennas Propag 63:1140–1149
4. Yang L, Li T, Yan S (2015) Highly compact MIMO antenna system for LTE/ISM applications. Int J Antennas Propag 2015:1–10
5. Nigam H, Kumar M (2014) Design and analysis of 2×2 MIMO system for 2.4 GHz ISM band applications. Int J Advance Research Compt Eng Technol 3:1794–1798
6. Anuvind R, Joseph SD (2015) 2×2 MIMO antenna at 2.4 GHz for WLAN applications. In: International Conference on Microwave, Optical and Communication Engineering, pp 80–83
7. Peristerianos A, Theopoulos A, Koutinos AG, Kaifas T, Siakavara K (2015) Dual-band fractal semi-printed element antenna arrays for MIMO applications. IEEE Antenna Wirel Propag Lett 15:730–733

8. Singh BK (2015) Design of rectangular microstrip patch antenna based on artificial neural network algorithm. In: International Conference on Signal Processing and Integrated Networks, pp 6–9
9. Arora P, Dhaliwal BS (2011) Parameter estimation of dual band elliptical fractal patch antenna using ANN. In: International Conference on Devices and Communications, pp 1–4
10. Dhaliwal BS, Pattnaik SS (2016) BFO–ANN ensemble hybrid algorithm to design compact fractal antenna for rectenna system. Neural Comput Applicat 28:1–12
11. Dhaliwal BS, Pattnaik SS (2017) Development of PSO-ANN ensemble hybrid algorithm and its application in compact crown circular fractal patch antenna design. Wirel Pers Commun 96:135–152
12. Kaur R, Josan SK, Dhaliwal BS (2017) Analysis of a pentagon fractal antenna using artificial neural network. In: International Conference on Soft Computing Applications in Wireless Communication, pp 214–218
13. Singh S, Dhaliwal BS (2017) Analysis of hybrid fractal antenna using artificial neural network. In: International Conference on Soft Computing Applications in Wireless Communication, pp 219–222

Chapter 16
Universal Approach for Detection of Spot Diseases in Plants

Aditya Sinha and Rajveer Singh Shekhawat

1 Introduction

Plants play an essential role in working and maintaining the balance on this earth. As humans have evolved and moved ahead in the technological era, we have generated a vast number of ways to feed the entire population. During the later phase of the twentieth century, we were able to sustain the entire human population food demands by respective technological growth [2] but this pressure might become unbearable for the entire civilization. As per the recent projections by Food and Agriculture Organization of the United Nations by 2050, we might have to feed 9.7 billion humans [3, 4] (Fig. 1) which means we have to produce 70% more food than what we are producing right now [5].

These projections present a very intense pressure on agriculture, which is still the primary source of food production for humans. Plants getting effected by pollution and by diseases make the situation worse. Some of the worst examples of the effects of plant diseases on the human population would draw our attention on some of the most horrific famines of the nineteenth century, i.e., Irish Potato Famine which happened due to potato blight disease and caused the death of nearly a million people [6]. When we look into the Bengal Famine of 1943 in India, a fungus namely Cochliobolus miyabeanus causing the brown spot disease in the rice was one of the significant reasons for this horrific tragedy. This famine caused the death of nearly 2 million people [7].

It is very much clear from the looking at these horrific tragedies how much effect the diseases can cause plant/crops. So it is of the urgent need to look into this matter with utmost priority. Now talking about plant diseases, they mainly can be categorized as biotic and abiotic. Plant diseases caused by the infectious/ parasitic pathogens

A. Sinha (✉) · R. Singh Shekhawat
School of Computing & Information Technology, Manipal University Jaipur, Rajasthan, India
e-mail: aditya.160301606@jaipur.manipal.edu

R. Singh Shekhawat
e-mail: rajveersingh.shekhawat@jaipur.manipal.edu

S. Jain et al. (eds.), *Advances in Computational Intelligence Techniques*, Algorithms for Intelligent Systems,
https://doi.org/10.1007/978-981-15-2620-6_16

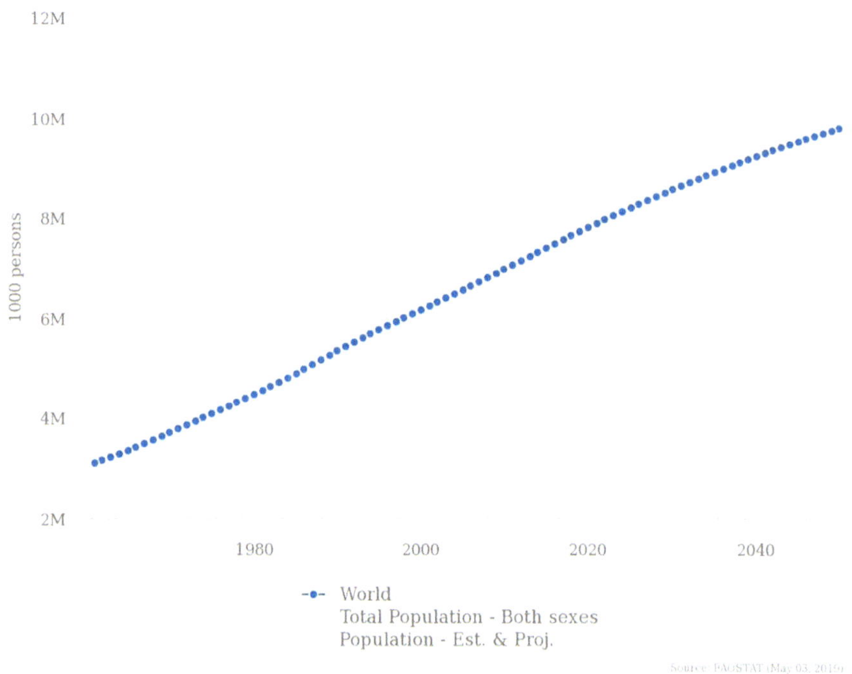

Fig. 1 World population projection [4]

are into biotic diseases, i.e., bacteria, fungi, virus, phytoplasma, and nematodes. Also, plant diseases which are non-infectious such as nutritional deficiencies hostile environment.

Conventionally, the plant disease detection was done at manual level; mainly the farmer used to inspect the crop/plant and asses the situation accordingly, as convenient it sounds, this process has its drawbacks as well. Since it requires a vast amount of experience to judge and asses the abnormalities in the crop/plant, this process is entirely subjective. There are frequent and obvious human errors also associated with this technique, such as tiredness, lack of concentration, and color blindness. Major drawbacks of a manual rater who evaluates the abnormality in the plant/crop are compiled by Bock et al. [8] as follows:

- Individuals vary in their intrinsic ability: Every individual has their unique way of assessing the situation, and hence, uniformity is not there in the inspection protocol.
- Value preferences by raters: Some raters prefer a specific value for the disease or its severity which is unique to their sense
- Lesion number and size relative to the area infected: How many numbers of lesions are there on the infected part of the plant may be subjective as there are many overlapping lesions.

- Actual disease severity: How much the infection is spread cannot be quantized into a specific value in case of manual raters as this can be subjective in the end.
- Time taken to assess disease: Time consumption in assessing each of the plant/crops in the whole field will always be more for manual raters.
- Color blindness: Some raters might be suffering from color blindness which hampers their ability to detect specific colors which might be an effect of the infection on the plant/crop.

Machine vision is used to solve the problems mentioned above; there has been much work in their field, and detection of the plant abnormality, i.e., plant disease detection, quantification, classification, and estimation, is a significant area of work in the field of image processing in recent years.

In this work, we have detected the leaf spot diseases in various plants. We have tested three main segmentation techniques for the detection, such as k-means clustering, LAB color histogram thresholding, and Delta-E color segmentation. K-means clustering with three clusters was the most efficient and generalized ROI approach for isolation of the infected area. Using the La*b* color space thresholding, we were also able to determine the infected area, but the threshold value was not fixed and was over a range.

2 Background

In [9], Lindow and Webb proposed a technique to quantify the severity of tomato, bracken fern disease, and California buckeye. They categorized the leaves into three categories according to the intensity values of the corresponding pixel, i.e., background, health, and necrotic area. They achieved 95% on confidence level in individual disease identification. Price et al. [10] devised a technique to measure the severity of the coffee rust on the coffee leaf. In their study, they found out that by visual estimation and overestimation of over two times occurred. Using their two digitizers, they were able to estimate the severity with more precision. They were also able to determine what shape, area, and perimeter of the affected area as well, which gave it a clear advantage over manual raters. Martin and Rybicki [11] developed a computer vision-based system to quantify the extent of damage caused by maize streak virus in maize crop. The authors used two commercially available digital image analysis systems; they measured the extent of severity by pixel intensity thresholding. The authors achieved better and more detailed results than visual raters. Škaloudová et al. [12] developed a technique to quantify the damage done by spider mites on the bean plant. They used the two-stage thresholding technique for disease estimation. They compared their results with chlorophyll fluorescence and leaf damage index method and got better results than those. Weizheng et al. [13] proposed a machine vision-based technique to grade soybean leaves affected by leaf spots. They used the Hue channel of the HSI color space to identify the infected area in the isolated area using edge detection technique, namely Sobel operators. Then according to the affected area and its extent leaves where graded. This approach gave a better understanding of the effects of damage on the soybean leaf due to leaf spot fungus. Medina et al. [14] devised and real-time computer vision technique based on FPGA, for the quantification of the common bean mosaic virus on the bean pepper and pumpkin leaves. They devised algorithms to identify chlorotic and necrotic area of the infected leaf. They also used pixel-based algorithm for the leaf deformation, white spot, and mosaic identification in the infected leaves. The whole system was based on an FPGA board, so it was real-time and rapid. Zhang et al. [15] proposed a sparse representation classification technique to identify mildew, scab, and anthracnose disease in cucumber leaves. They reduced the number of features using the sparse representation technique. After diseased area segmentation using k-means, they reduced the computation cost using the sparse representation. They achieved an accuracy of 85.7%.

3 Analysis of Spot Diseases

In this research work, we have worked on detection of spot diseases affecting various plants. We have looked into diseases, namely maize Cercospora leaf spot, tomato Septoria leaf spot, bell pepper bacterial spot, grape Isariopsis spot Neofabraea leaf spot, and tomato bacteria leaf spot. Neofabraea leaf spot occurs when the leaves are damaged due to harvesting, in the damaged area of the leaf the fungus grows, and symptoms appear on the leaf, trunk, and stem. On the leaf, there is a clear distinction in color, the infected area is brown, and they usually look like a spot. Images are collected from an online source; we have collected mostly the images of intermediate damage stage on olive leaves. Images collected from various verified online sources. Images collected and used in the experiment for the detection of Neofabraea are numbered as Neofabraea (NB) followed by sequential numbering, a sample can be seen in Fig. 2.

Also the whole dataset as collected from the PlantVillage for the other diseases can be seen in Fig. 3.

For the experimentation images are transformed form Red, Green, Blue (RGB) color model to La*b*(L* for the lightness from black to white, a* from green to red, and b* from blue to yellow.), HSV (Hue, Saturation, Value), and YCbCr (luma component, blue-difference, and red-difference chroma components) color models, out of these on after evaluation and experimentation La*b* color model was selected. The threshold value on a* and b* channel segmented the diseased region properly. We can see the histogram values of the La*b* color model of sample NB1 leaf in Fig. 4. After experimentation, the maximum threshold value of the a* channel was found to be in the 20–32 range, and minimum value for channel a* was found out to be in the range of 1 to −9. After checking in the LAB color space and especially in the a* channel, the infection was prominent and quite visible in comparison with other color models.

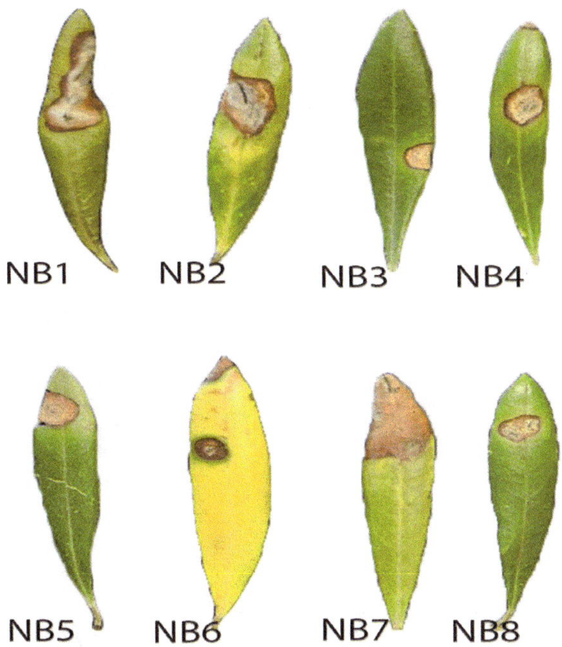

Fig. 2 Collection of olive leaf images affected by Neofabraea Blight

Fig. 3 Dataset

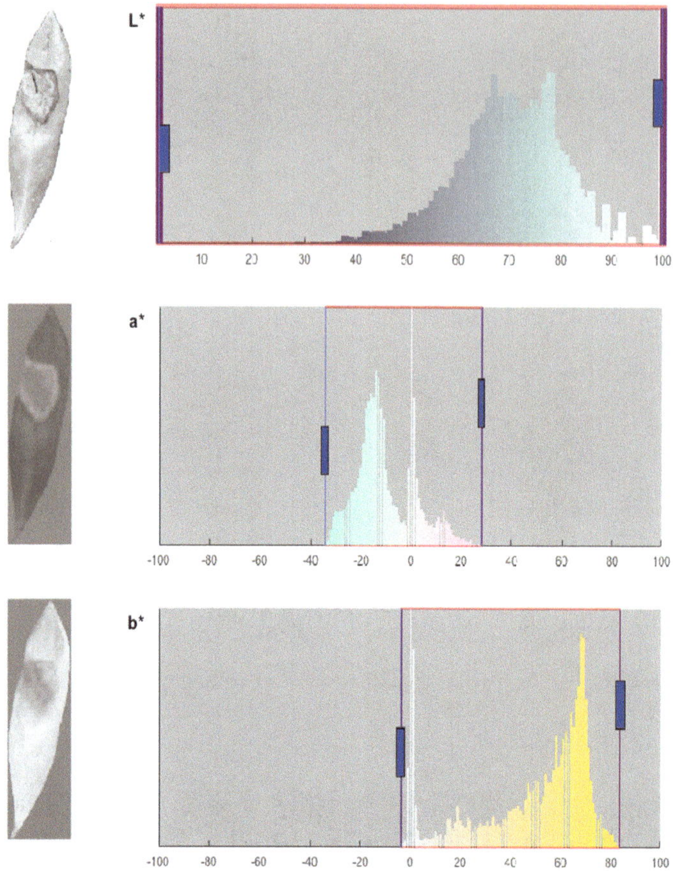

Fig. 4 Histogram of olive leaf in La*b* color space affected by Neofabraea Blight

Distribution of the threshold as applied on the minimum and maximum values for a* and b* channel of La*b* color model histogram for each leaf sample affected by Neofabraea disease can be seen in Fig. 5.

The results obtained were giving the expected results, and the leaf area, which was infected, was isolated properly by this thresholding technique. We also tried the k-means clustering for infected area segmentation, mainly to isolate the region of interest (ROI). We obtained a satisfactory result in the same. Results obtained by both the technique can be seen in the result section. There were three clusters for the primary three dominating colors found in the infected leaf samples, mostly infected color, i.e., reddish-brown was appropriately isolated using this technique. k-means clustering holds and advantages over the thresholding technique in this experimentation as we had to look into various infected leaf samples for the thresholding values to work.

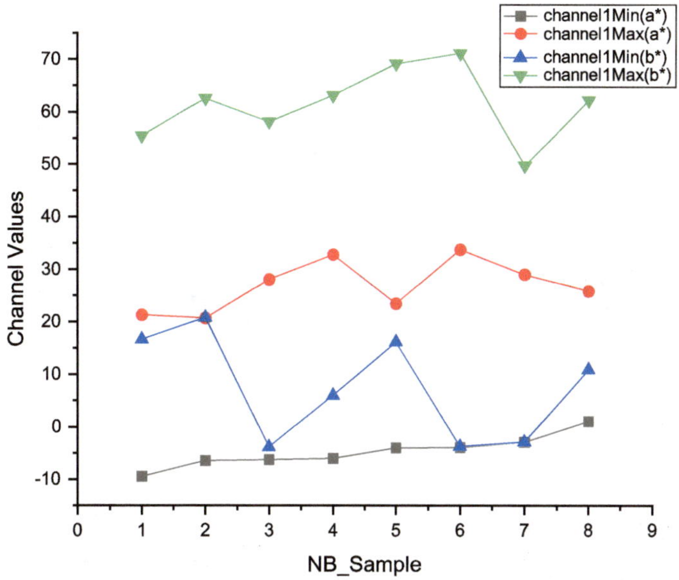

Fig. 5 Minimum and maximum threshold values threshold for b* & a* channel of olive leaf images affected by Neofabraea blight

4 Results

In this section, we look into the results obtained by applying all the three techniques mentioned above, i.e., thresholding on the La*b* color histogram values, k-means clustering-based segmentation method, and the Delta-E color segmentation on the dataset of leaf spot images. In Fig. 6, we can see the application of the threshold values on some leaf samples affected by the pathogen as we can see almost all the infected region of interest (ROI). Applying the threshold values mainly on the minimum a* channel gave the perfect results, and the infected regions were properly isolated.

Results obtained after applying k-means clustering for clustering dividing pixels in three clusters can be seen in Fig. 7. While segmenting the ROI using k-means, we tried different cluster numbers and different centroid positions. In general for the whole dataset, 3 cluster-based k-means worked the best and was able to isolate most of the time. There was some infected area which was not properly isolated, but this will be fixed in the future work of this experimentation.

Some of the leaf samples were already effected by chronic chlorine deficiency, and hence, in some cases, the algorithm did not work properly. Also the results obtained on disease dataset through Delta-E color segmentation and LAB color thresholding can be seen in Figs. 9 and 10. While using the Delta-E color segmentation, the Delta-E value was varying most of the time for each of the diseases but was quite

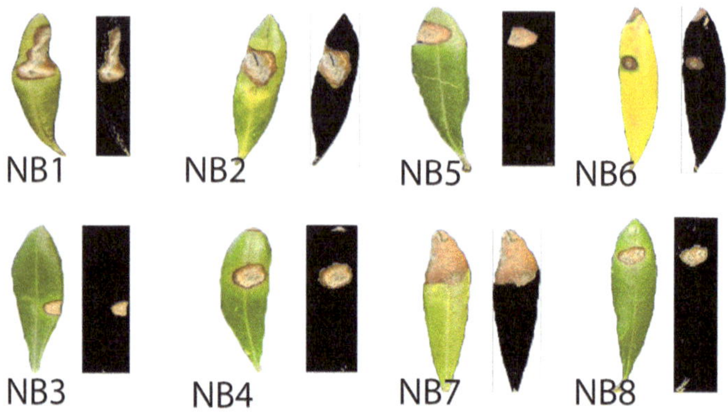

Fig. 6 Output of images after applying threshold on a* channel of La*b* histogram values on olive leaf affected by Neofabraea leaf spot

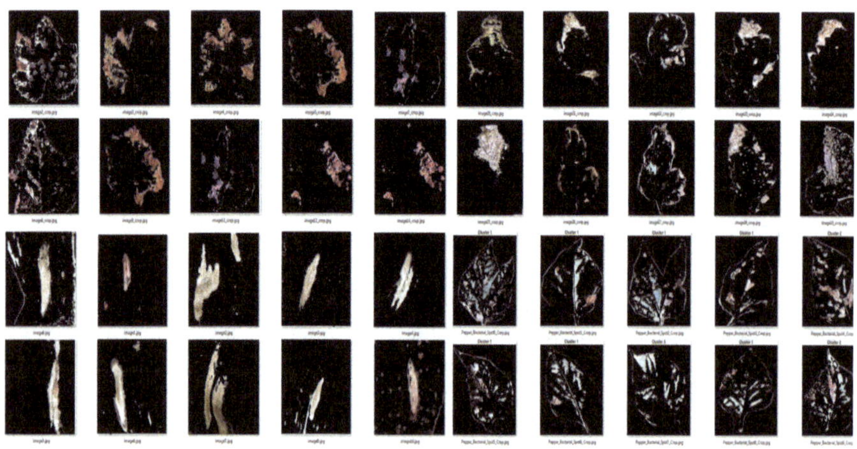

Fig. 7 Output of images after applying k-means clustering on leaf affected by spot diseases

similar in the same class of diseases. LAB color thresholding was able to isolate the ROI properly, but it was hard to get upper and lower bounds of the threshold which can be applied over the whole dataset and hence making the whole process not general.

We can see the comparison in the infected area (%) according to both techniques in the corresponding image samples in Fig. 8.

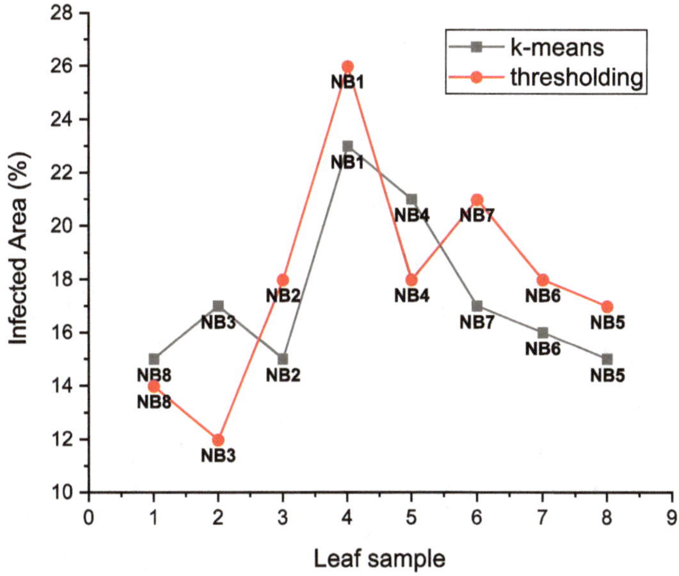

Fig. 8 Infected are comparison between both techniques

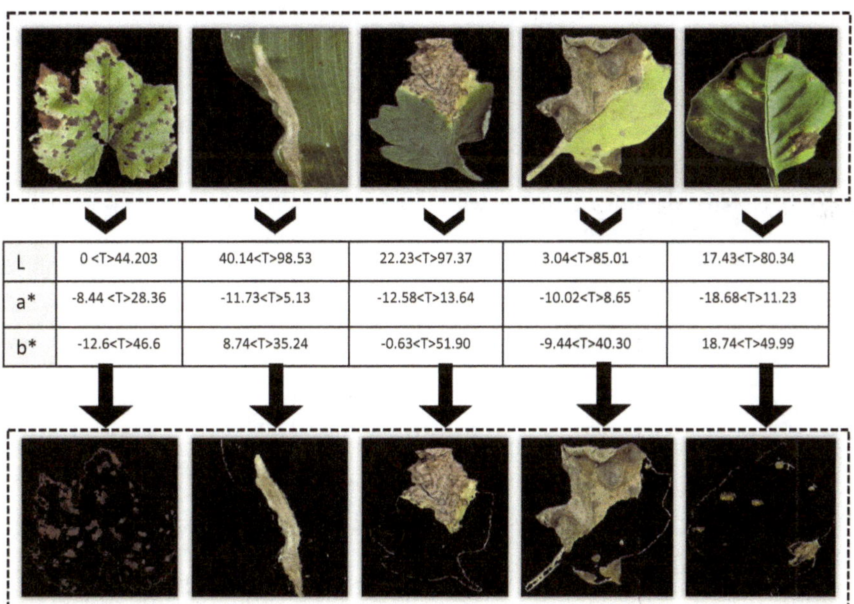

L	0 <T>44.203	40.14<T>98.53	22.23<T>97.37	3.04<T>85.01	17.43<T>80.34
a*	-8.44 <T>28.36	-11.73<T>5.13	-12.58<T>13.64	-10.02<T>8.65	-18.68<T>11.23
b*	-12.6<T>46.6	8.74<T>35.24	-0.63<T>51.90	-9.44<T>40.30	18.74<T>49.99

Fig. 9 LabROI

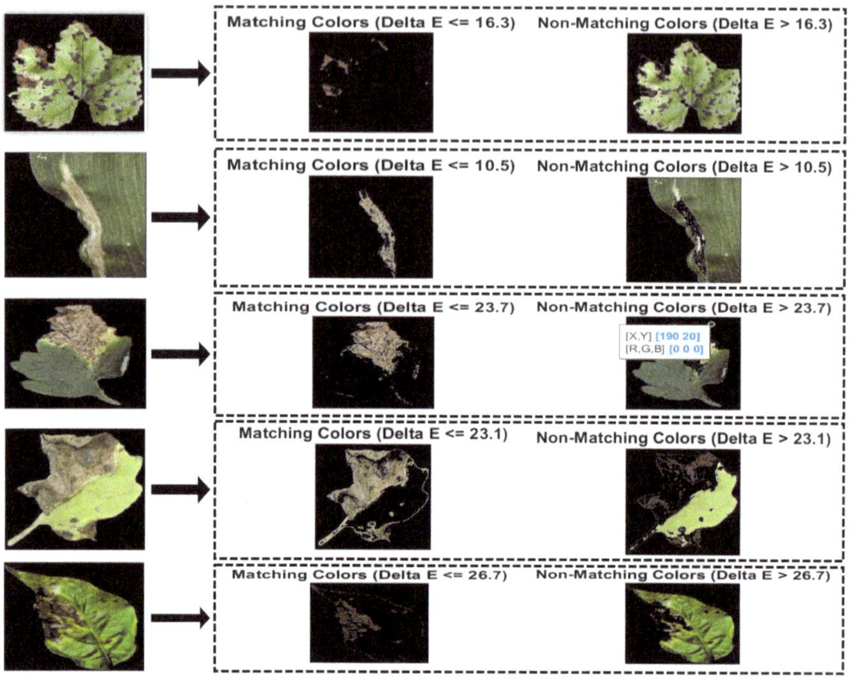

Fig. 10 Delta-E color ROI

5 Future Work

We have isolated the infected area caused by spot diseases. Also, the expansion of the dataset is potentially future work. This technique will be tested on other olives diseases as well as on similar visual symptoms base plants/crops diseases. Some public dataset such as [16, 17] will be used for testing these techniques on various other diseases and crop types.

6 Conclusion

We tested three different techniques, i.e., Delta-E color segmentation, thresholding, and k-means classification for the identification and isolation of the infected area of leaf spot diseases in various plants. La*b* color thresholding model proved to be most accurate when identifying the infected area, but using k-means classification over the whole dataset gave the most generalized results overall (Fig. 8). It was not possible to get a fixed bound in Delta-E and thresholding, and hence, generalization was very hard. These techniques were tested on a small dataset, but gave good accuracy and will be expanded on a larger dataset as well. We have worked on various spot diseases;

we will be expanding this work on other pathogen-based diseases. For the detection, a novel algorithm is devised to isolate the infected region more prominently. The validation of the results and the technique will also occur with testing the technique on a more elaborate dataset. For classification, a novel algorithm will give more confidence to the results.

Acknowledgements Authors are thankful to Manipal University Jaipur for all the support in the research work. Thanks are also due to the head of the department, Computer Science & Engineering and other staff for their help and interest.

References

1. Arnal Barbedo JG (2013) Digital image processing techniques for detecting, quantifying and classifying plant diseases. SpringerPlus 2:660
2. Keating B, Carberry P (2010) Sustainable production, food security and supply chain implications
3. Keating BA, Herrero M, Carberry PS, Gardner J, Cole MB (2014) Food wedges: framing the global food demand and supply challenge towards 2050. Global Food Secur 3(3):125–132 (SI: GFS Conference 2013. [Online]. Available at http://www.sciencedirect.com/science/article/pii/S2211912414000327)
4. Faostat: Fao statistical database. Retrieved. 2019, May [Online]. Available at http://www.fao.org/faostat/en/
5. Cole MB, Augustin MA, Robertson MJ, Manners JM (2018) The science of food security. npj Sci Food 2(1). https://doi.org/10.1038/s41538-018-0021-9
6. Kinealy C (1994) This great calamity: the great Irish Famine: the Irish Famine 1845-52. Gill & Macmillan Ltd
7. Bengal famine of 1943. Accessed on May 2019. [Online]. Available at https://archive.org/stream/in.ernet.dli.2015.206311/2015.206311.Famine-Inquirypage/n41/mode/2up/search/fungus
8. Bock C, Poole G, Parker P, Gottwald T (2010) Plant disease severity estimated visually, by digital photography and image analysis, and by hyperspectral imaging. Crit Rev Plant Sci 29(2):59–107
9. Lindow S, Webb R (1983) Quantification of foliar plant disease symptoms by microcomputer-digitized video image analysis. Phytopathology 73(4):520–524
10. Price T, Gross R, Ho WJ, Osborne C (1993) A comparison of visual and digital image-processing methods in quantifying the severity of coffee leaf rust (*Hemileia vastatrix*). Aust J Exp Agric 33(1):97–101
11. Martin DP, Rybicki EP (1998) Microcomputer-based quantification of maize streak virus symptoms in zea mays. Phytopathology 88(5):422–427
12. Škaloudová B, Křivan V, Zemek R (2006) Computer-assisted estimation of leaf damage caused by spider mites. Comput Electron Agric 53(2):81–91
13. Weizheng S, Yachun W, Zhanliang C, Hongda W (2008) Grading method of leaf spot disease based on image processing. In: 2008 international conference on computer science and software engineering. IEEE, New York, pp 491–494
14. Contreras-Medina LM, Osornio-Rios RA, Torres-Pacheco I, de J Romero-Troncoso R, Guevara-González RG, Millan-Almaraz JR (2012) Smart sensor for real-time quantification of common symptoms present in unhealthy plants. Sensors 12(1):784–805
15. Zhang S, Wu X, You Z, Zhang L (2017) Leaf image based cucumber disease recognition using sparse representation classification. Comput Electron Agric 134:135–141

16. Plantix plant disease libary (2018) Available at https://plantix.net/plant-disease/en
17. Leafsnap dataset (2018) Available at http://leafsnap.com/dataset/
18. PlantVillage dataset. Accessed 1 Nov 2018. Available at https://github.com/spMohanty/PlantVillage-Dataset/tree/master/raw/color/

Chapter 17
Stereo Camera and LIDAR Sensor Fusion-Based Collision Warning System for Autonomous Vehicles

Amara Dinesh Kumar, R. Karthika and K. P. Soman

1 Introduction

Automobile has become an integral part of today's modern society. Recent advances in the computer vision and automotive embedded systems have accelerated the automation in automobiles. Autonomous ground vehicles are the next generation cars with the capabilities of autonomous driving. The autonomous cars have different components like perception, localization, navigation and control systems. These systems have to constantly communicate and exchange information from each other and work in integrated environment for decision-making in real time. Perception system is the primary and most complicated system which senses the environment through different sensors then processes it and sends the information to the other systems.

Throughout history, many attempts were carried out to develop an autonomous vehicle that can travel without any human driver like the Leonardo da Vinci's self-propelled cart and Dickmann's VaMP autonomous vehicle using its 4D vision technology. But the huge leap was the initiative of Autonomous Land Vehicle (ALV) project from Defense Advanced Research Projects Agency (DARPA) which uses sophisticated sensors and capable of navigating autonomously at high speeds. DARPA has started a series of challenges from the 2004 to 2007 period. Those challenges triggered the research that pushed the capabilities of autonomous vehicles. After that Google has started developing its own self-driving car and soon other companies like Telsa, Uber and Nvidia have joined the race.

A. Dinesh Kumar (✉) · R. Karthika (✉)
Department of Electronics and Communication Engineering, Amrita School of Engineering, Amrita Vishwa Vidyapeetham, Coimbatore, India
e-mail: r_karthika@cb.amrita.edu

K. P. Soman
Centre for Computational Engineering and Networking (CEN), Amrita School of Engineering, Amrita Vishwa Vidyapeetham, Coimbatore, India

S. Jain et al. (eds.), *Advances in Computational Intelligence Techniques*, Algorithms for Intelligent Systems,
https://doi.org/10.1007/978-981-15-2620-6_17

Our main contributions from this research are

1. To detect and identify the obstacles in front of the vehicle by fusing the LIDAR data and camera data and developing a forward collision warning system.
2. Collecting and synchronizing the distance data from the LIDAR and image data from camera simultaneously in real time.
3. Calibrating both camera and LIDAR when mounted onto real vehicle by processing the data of both camera and LIDAR sensors and then performing point to pixel data fusion.
4. Simultaneously detecting and classifying obstacles (car, pedestrian, etc.) using deep learning algorithm.

The paper is organized as follows: Sect. 2 includes the related work. Section 3 discusses different vehicle detection and localization algorithms, Sect. 4 describes about the proposed methodology, Sect. 5 contains the experimental analysis and the results, and Sect. 6 contains conclusion.

2 Related Work

Real-time obstacle detection and classification are challenging tasks for advanced driver assistance systems, robot navigation and surveillance [1]. Conventionally obstacle detection and classification were tackled using RADAR and LIDAR sensors, but because of the high cost and poor ability in estimating the size and the shape of vehicles, camera-related solutions have recently gained attention in autonomous vehicles research community. Availability of large datasets [2–4], embedded processing power [5], led to an interest in applying deep learning architectures for real-time detection, localization and classification of the obstacles, and they performed superior to the prior algorithms recently [6–8].

Extensive research was done using conventional image processing techniques and manual feature extraction methods for object detection and localization. Histogram of oriented gradients (HOG) and scale-invariant feature transform (SIFT) are generally used for feature extraction from the target image along with support vector machine (SVM) learning algorithm is commonly used [8]. Different convolutional neural networks' architectures were proposed for real-time object detection and localization and widely used are You look only once (YOLO), fast region convolutional network (RCNN), faster RCNN, tiny YOLO and SSD [9, 10].

Popular track by detection algorithms uses discriminative appearance-based model which consists of a classifier that classifies object presence in an image patch [11]. Supervised learning algorithms in machine learning like SVM, Random forest [12], Naive Bayes classifier and other boosting algorithms [13–16] are widely used in predicting the target in the image patch [17–20]. But the machine learning algorithms need offline prior training with a training dataset [21–24], but the biggest disadvantage was they perform well only for the testing data that is similar to training data [8, 25–27].

Scaramuzza et al. [28] used Haar with AdaBoost for vehicle detection and SVM algorithm for vehicle classification with 91.5 accuracy. Shape estimation is performed with random hypersurface model. Feng et al. [29] used the RADAR and camera sensors and performed the sensor fusion along with lane detection and segmentation; Kalman filter was used for sensor fusion. Hopkins [30] used the HOG for feature extraction and SVM for vehicle detection and classification. Kernelized correlation filter (KCF) tracker is used for tracking with a prime sense camera and RPLIDAR 360°. Kalman filter and its variants are the widely used state estimation algorithms for the sensor fusion of the camera and LIDAR sensors in the literature [29, 30].

2.1 *Sensors Used for Development of Autonomous Vehicle Perception System*

LIDAR refers to light detection and ranging. It uses light pulses instead of the electromagnetic waves for detecting the ranges. It operates with almost the same principles of RADAR. They are different types of LIDARs available like 3D and spinning 2D (Fig. 1).

LIDAR sensors provide high precision, accuracy in object detection and recognition in ADAS systems by providing millions of data points per second creating a 3D point cloud reconstruction of the surrounding (Fig. 2).

LIDAR gives positional accuracy and precision with high data rate. Major disadvantage of LIDARs is bulkiness and cost for using them for the ADAS applications. The high-resolution 3D LIDAR costs very high, and with the advances in the LIDAR technology, it may decrease in the future.

Stereo camera sensor is a passive reflective device with a combination of two cameras. It can be used for 3D reconstruction of the environment eliminating the requirement for high-cost LIDAR (Fig. 3).

The disparity between the left and right camera images was calculated then from the disparity map. The relative distance of an object from the stereo camera within the field of view can be calculated.

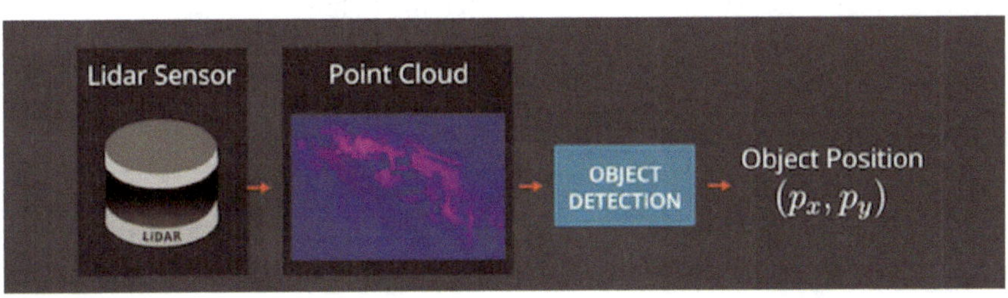

Fig. 1 Object detection using LIDAR

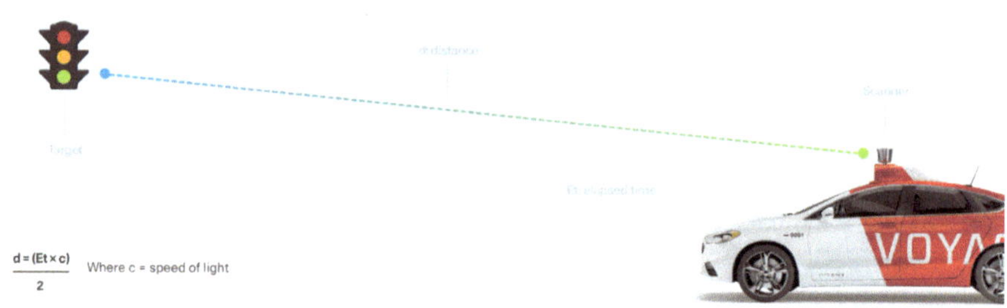

Fig. 2 Illustration of LIDAR working in autonomous vehicles

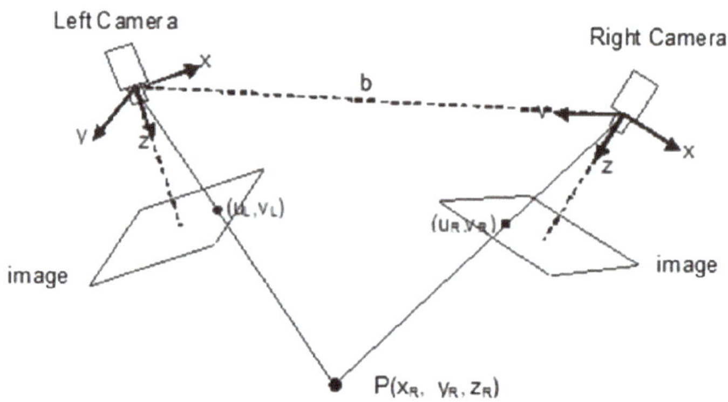

Fig. 3 Working principle of stereo vision camera

The depth is calculated as

$$\text{Depth} = f \times b/d \quad (1)$$

where b = left camera and right camera distance, d = disparity, f = focal length.

The U_L and U_R are the projected points distance on the image plane. The major advantage of stereo camera sensor is along with camera functionality; it can also provide the obstacles' distance. Binocular stereo vision system has challenges as an inaccurate depth estimation in extreme weather and lighting conditions.

3 Detection and Localization Algorithms

In embedded and real-time applications, MobileNet SSD Architecture has widely used for mobile robots and autonomous vehicles perception tasks because of its streamlined architecture and lightweight (use of depth-wise separable convolutions).

It is a combination of SSD (used for object detection and localization) and MobileNet (used for object classification).

3.1 *MobileNet*

MobileNets are network architectures constructed for the purpose of running efficiently on mobile embedded-based vision application. MobileNets achieve this with the following techniques:

- Rather than a standard convolution perform depth-wise convolution followed by a 1×1 convolution. If 1×1 convolution is following a depth-wise convolution, then it is called a point-wise convolution. A separable depth-wise convolution is the combination of depth-wise convolution followed by a point-wise convolution.
- Reducing the size of input/output channels by using a width multiplier, set to a value between 0 and 1.
- Reducing the size of original input using a width multiplier, set to a value between 0 and 1.

The size of cumulative parameters and the computation required can be reduced using the above techniques. Usually, models produce higher accuracy and were optimized for mobile devices.

3.2 *Single-Shot Multibox Detection (SSD)*

Single-shot detection (SSD) is a popular algorithm in object detection which is faster than region convolution neural network (RCNN) and region proposal network (RPN). The detailed architecture and layers are shown in Fig. 4.

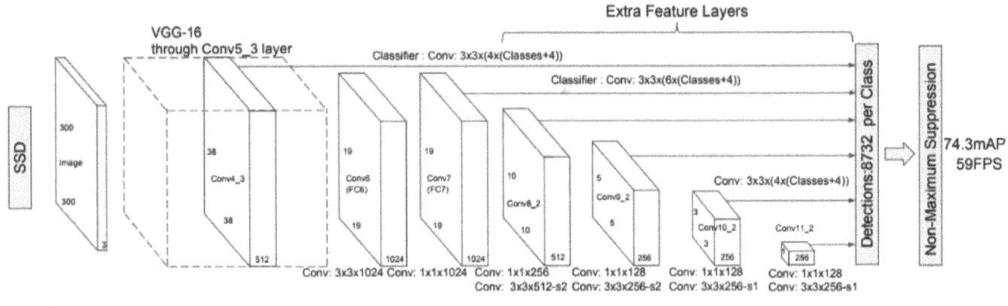

Fig. 4 Single-shot multibox detector (SSD) architecture [31]

4 Proposed Methodology

The stereo camera initially produces both the left images and the right images. After that rectification of both images is performed. The rectification is a process of computing the image transformation which converts epipolar lines collinear that are parallel to horizontal axis given the extrinsic parameters of the systems and intrinsic parameters of the camera from the rectified images; the disparity map is calculated by utilizing the disparity map; the pixel-wise depth can be calculated (Fig. 5).

Obstacle detection is performed from the left image using the pre-trained MobileNet single-shot multibox detector (SSD) algorithm. The detailed architecture of MobileNet SSD algorithm is showed. It is trained to detect total six classes, and they are pedestrian, vehicle, bicyclist, cow, bus and dog classes. Coordinates of the bounding box of the detected obstacle are calculated, and then the obstacle's distance is calculated as

$$d = f \times b/m \tag{2}$$

f = focal length, d = calculated distance of stereo camera, b = baseline, m = mean of the disparity in bounding box region

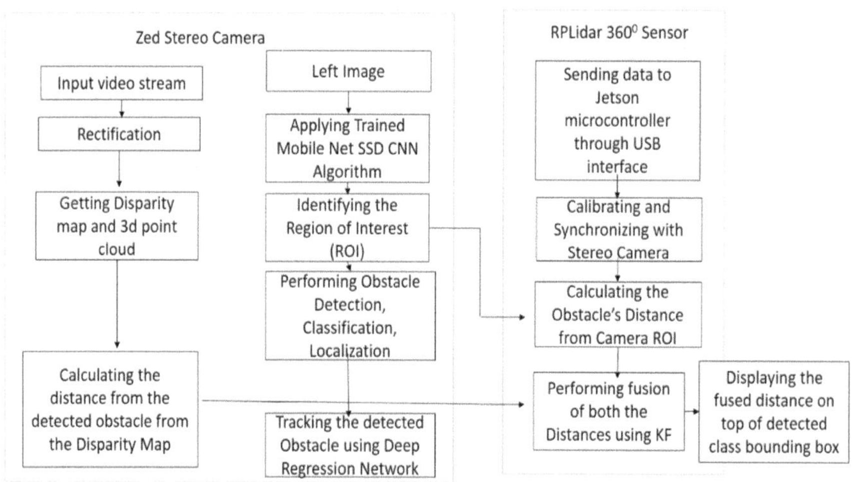

Fig. 5 Block diagram for stereo camera and RPLidar 360° sensor-based collision warning system

Then, the corresponding start angle and end angles are calculated for calibrating and synchronizing with the LIDAR distance. The stereo camera angles are transformed to match with LIDAR angles for equivalent horizontal field of view and mapped correspondingly using the below algorithm.

Algorithm 1:

deg = h fov/w, (that implies 1 pixel = 86.9 / 672 = 0.1293 degree s)
midp = w / 2 (= 336 s)
angs = startX ∗ deg
ange = endX ∗ deg

if startX is less than the mid point
dif = midp − startX
angs = 360 − dif ∗ deg
else
dif = startX − midp
angs = dif ∗ deg

h fov = horizontal field of view , w = width of the image, angs = start angle, ange = end angle, midp = midpoint, deg = degree,

Now the LIDAR distance is calculated by taking the mean of the distances for corresponding angles of LIDAR horizontal field of view according to the sampling rate.

4.1 Sensor Calibration

The ZED camera calibration consists of intrinsic parameters for both left and right sensors at each resolution. Figure 6 shows the left and right images from the stereo camera along with the corresponding depth images generated (Table 1).

fx and fy are the focal length in pixels. cx and cy are the optical center coordinates in pixels, and k1 and k2 are distortion parameters (Fig. 7).

The stereo parameters, also called extrinsic parameters, represent the relation between the left and right sensors. More precisely, the position of the right sensor from the left sensor with the center of rotation is the right sensor itself (Table 2).

Baseline is the distance between the optics in mm (120 for the ZED and 63 for the ZED Mini); CV also called RY measures the optical convergence. RX and RZ are the other rotation axes describing the transformation between both sensors. Rotations are represented in the Rodrigues notation.

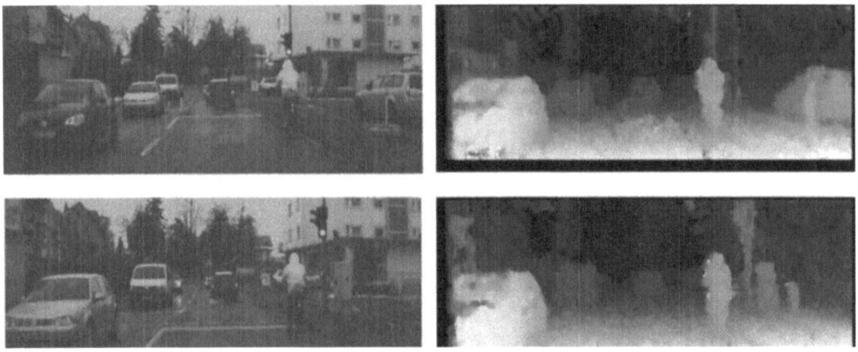

Fig. 6 Stereo vision images along with corresponding disparity map [32]

Table 1 Left camera parameters

Parameter	Value
fx	700.819
fy	700.819
cx	665.465
cy	371.953
k1	−0.174318
k2	0.0261121

Fig. 7 Sensors setup and placement

Table 2 Stereo parameters

Parameter	Value
Baseline	120
CV HD	0.00958521
RX HD	0.00497864
RZ HD	−0.00185401

4.2 Sensor Synchronization

Stereo camera is operated at 60 frames per second (FPS) with resolution of 1280 × 720 pixels. The offset between the stereo camera and LIDAR is taken as 200 mm. The synchronous centralized data fusion is performed by using a sampling rate of 1 s. After every sampling instant, the distances from both the sensors are processed and sent to the sensor fusion algorithm for fusing both the values.

5 Experimental Analysis and Results

5.1 Stereo Camera and RPLIDAR Sensor Data Fusion

The true distance is measured and marked for comparing with the fused distance. A stationary pedestrian was used for taking the measurements for the experiment (Fig. 8).

The readings from 1000 to 5500 mm distance were recorded for the stereo camera along with variance, and LIDAR sensor with variance is measured by periodically moving the pedestrian subsequently in 500 mm intervals and shown in Table 3. Kalman filter algorithm is applied, and the measured distances of the two sensors along with their variances are combined together resulting in a new distance measurement with lesser variance and uncertainty.

The final result consists of the distance calculated from the stereo camera and LIDAR sensor along with the fused distance value. The obstacle is detected and shown in the image with obtained bounding box coordinates and classified into the respective class that is displayed on top of the bounding box along with the confidence score as shown in Fig. 9. The camera distance and LIDAR distance are displayed along with the fused distance above the bounding box. The detected class (pedestrian) along with confidence score is also displayed.

The performance of the sensor fusion is evaluated using the RMS and MAE standard error metrics. The RMS error of the stereo camera, LIDAR sensor and the final fused distance are calculated. The RMS error has significantly decreased for the fused distances making it reliable and accurate compared to the individual sensor measurements.

Fig. 8 Pedestrian standing at the marking

Table 3 Kalman filter fused distance and variance

Distance (mm)	Camera distance	Camera variance	LIDAR distance	LIDAR variance
1000	1020.681	74.455	1117	105.193
1500	1521.944	0.0946	1637.75	31.813
2000	2081.26	3.413	2049.25	18.165
2500	2540.351	10.147	2621.25	47.792
3000	3057.945	0.036	3093.5	36.851
3500	3519.458	4.039	3632.25	14.995
4000	4031.859	4.927	4156	36.666
4500	4639.656	0.356	4636.625	24.096351
5000	5558.287	870.923	991.25	55.26
5500	5848.671	13776.96	936.25	158.4427

$$\text{Root Mean Squared Error (RMSE)} = \sqrt{\frac{1}{n}\sum\nolimits_{j=1}^{n}\left(y_j - \hat{y}_j\right)^2} \tag{3}$$

y_j is the measured distance and $\hat{y}_j$ is the actual (ground truth) distance (Table 4).

The mean absolute error (MAE) was also calculated for the comparison, and with 83.453 mm, the fused distance has lower error compared to the other sensor measurements (Table 5).

Fig. 9 Result of stereo camera and RP LIDARs' distance fusion

Table 4 Comparison of root mean squared error (RMSE)

Sensor type	RMS error (in mm)
Stereo camera	216.021180747
LIDAR sensor	1923.9780798
Fused distance	93.8027394119

Table 5 Comparison of mean absolute error (MAE)

Sensor type	MAE error (in mm)
Stereo camera	132.0112
LIDAR sensor	951.6125
Fused distance	83.45388

$$\text{Mean Absolute Error (MAE)} = \frac{1}{n}\sum_{t=1}^{n}\left|y_j - \hat{y}_j\right| \tag{4}$$

Figure 10 shows the comparison between the resultant fused distance and the original distance. The fused distance measurements obtained after fusing both the camera measurements and LIDAR sensor measurements were much closer compared to the individual measurements.

Figure 11 shows the comparison between the individual stereo camera measured distance, LIDAR sensor measured distance and the resultant fused distance obtained from the developed fusion algorithm.

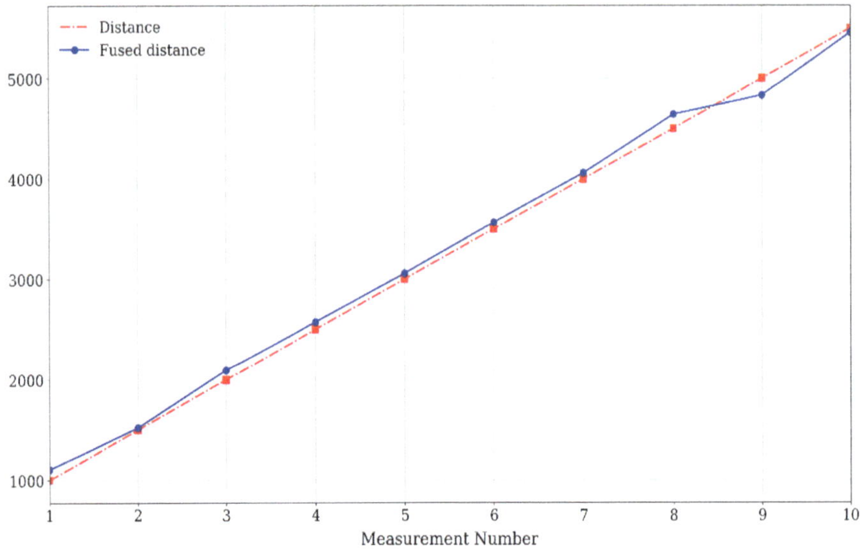

Fig. 10 Comparing fused distance with original distance

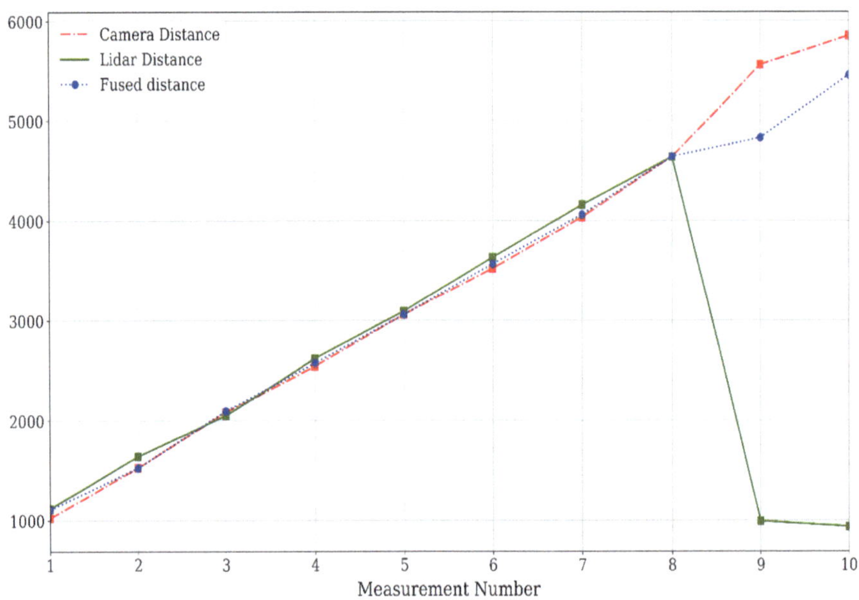

Fig. 11 Comparing fused distance with camera and LIDAR distance

6 Conclusion

A forward collision warning system is developed and evaluated using the 2D LIDAR and stereo camera sensors that work in real time on an embedded board. Deep learning-based MobileNet SSD algorithm is used for detection, classification and localization of the obstacles coming in front of the moving vehicle. The distances measured from both the sensors were fused with Kalman filter algorithm performing sensor fusion. The results signify that combining the distance measurements obtained from both the sensors by applying the proposed fusion algorithm decreased the variance and uncertainty and increased the reliability in the measurements. Sensor fusion combines the advantages of both 2D LIDAR and stereo camera sensors which make it more robust and immune to adversarial conditions. The developed system by fusing and complimenting data from both the sensors can work in extreme weather condition where individual sensors face challenges in perception.

References

1. Geronimo D et al (2010) Survey of pedestrian detection for advanced driver assistance systems. IEEE Trans Pattern Anal Mach Intell 32(7):1239–1258
2. Geiger A et al (2013) Vision meets robotics: the KITTI dataset. Int J Robot Res 32(11):1231–1237
3. Geiger A, Lenz P, Urtasun R (2012) Are we ready for autonomous driving? The kitti vision benchmark suite. In: 2012 IEEE conference on computer vision and pattern recognition (CVPR). IEEE
4. Cordts M et al (2016) The cityscapes dataset for semantic urban scene understanding. In: Proceedings of the IEEE conference on computer vision and pattern recognition
5. Meyer GG, Främling K, Holmström J (2009) Intelligent products: a survey. Comput. Ind. 60(3):137–148
6. Udalski A (2004) The optical gravitational lensing experiment. Real time data analysis systems in the OGLE-III survey. arXiv preprint astro-ph/0401123
7. Leonard JJ, Durrant-Whyte HF (1991) Mobile robot localization by tracking geometric beacons. IEEE Trans Robot Autom 7(3):376–382
8. Deepika N, Sajith Variyar VV (2017) Obstacle classification and detection for vision based navigation for autonomous driving. In: 2017 international conference on advances in computing, communications and informatics (ICACCI). IEEE
9. Lienhart R (2001) Reliable transition detection in videos: a survey and practitioner's guide. Int J Image Graph 1(03):469–486
10. Ma C et al (2015) Hierarchical convolutional features for visual tracking. In: Proceedings of the IEEE international conference on computer vision
11. Li X (2013) A survey of appearance models in visual object tracking. ACM Trans Intell Syst Technol (TIST) 4(4):58
12. Kulkarni VY, Sinha PK (2013) Random forest classifiers: a survey and future research directions. Int J Adv Comput 36(1):1144–1153
13. Li X (2013) A survey of appearance models in visual object tracking. ACM Trans Intell Syst Technol (TIST) 4(4): 58
14. Li Y, Zhu J (2014) A scale adaptive kernel correlation filter tracker with feature integration. In: European conference on computer vision. Springer, Cham

15. Smeulders AWM, Chu DM, Cucchiara R, Calderara S, Dehghan A, Shah M (2013) Visual tracking: an experimental survey. IEEE Trans Pattern Anal Mach Intell 36(7): 1442
16. Yang H, Shao L, Zheng F, Wang L, Song Z (2011) Recent advances and trends in visual tracking: a review. Neurocomputing 74:3823–3831
17. Zhang K, Zhang L, Yang MH (2012) Real-time compressive tracking. In: European conference on computer vision, Springer, pp 864–877
18. Kalal Z, Mikolajczyk K, Matas J et al (2012) Tracking-learning-detection. IEEE Trans Pattern Anal Mach Intell 34:1409
19. Babenko B, Yang MH, Belongie S (2011) Robust object tracking with online multiple instance learning. IEEE Trans Pattern Anal Mach Intell 33: 1619–1632
20. Saffari A, Leistner C, Santner J, Godec M, Bischof H (2009) On-line random forests. In: 2009 IEEE 12th international conference on computer vision workshops (ICCV Workshops), IEEE, pp 1393–1400
21. Grabner H, Grabner M, Bischof H (2006) Real-time tracking via on-line boosting. In Bmvc, vol 1, p 6
22. Grabner H, Bischof H (2006) On-line boosting and vision. IEEE
23. Henriques JF, Caseiro R, Martins P, Batista J (2015) High-speed tracking with kernelized correlation filters. IEEE Trans Pattern Anal Mach Intell 37:583–596
24. Kalal Z, Mikolajczyk K, Matas J (2010) Forward-backward error: automatic detection of tracking failures. In: 2010 20th international conference on pattern recognition (ICPR), IEEE, pp 2756–2759
25. Held D, Thrun S, Savarese S (2016) Learning to track at 100 fps with deep regression networks. In: European conference on computer vision, Springer, Cham
26. Jose A, Thodupunoori H, Nair BB (2019) A novel traffic sign recognition system combining Viola–Jones framework and deep learning. In: Soft computing and signal processing, Springer, Singapore, pp 507–517
27. Scaramuzza D, Achtelik MC, Doitsidis L, Friedrich F, Kosmatopoulos E, Martinelli A, Gurdan D (2014) Vision-controlled micro flying robots: from system design to autonomous navigation and mapping in GPS-denied environments. IEEE Robot Autom Mag 21(3):26–40
28. Zhang F, Clarke D, Knoll A (2014) Vehicle detection based on lidar and camera fusion. In 17th international IEEE conference on intelligent transportation systems (ITSC), October, IEEE, pp 1620–1625
29. Feng Y, Pickering S, Chappell E, Iravani P, Brace C (2017) Distance estimation by fusing radar and monocular camera with Kalman filter (No. 2017-01-1978). SAE Technical Paper
30. Hopkins AI. DJI RoboMaster AI challenge. Technical Report
31. Liu W et al (2016) Ssd: single shot multibox detector. In: European conference on computer vision, Springer, Cham
32. Darms M et al (2010) Data fusion strategies in advanced driver assistance systems. In: SAE international journal of passenger cars-electronic and electrical systems 3.2010-01-2337, pp 176–182

Chapter 18
Support Vector Machine-Based Direction of Arrival Estimation with Uniform Linear Array

Shardul Yadav, Mohd Wajid and Mohammed Usman

1 Introduction

Direction of arrival (DOA) estimation of an acoustic source has wide range of applications such as robotic movement in unknown environment, underwater/aerial target surveillance and tracking, underwater acoustic vector sensor communication, automatic steering of camera towards speaker in a room, room speech enhancement, smart home automation, hearing aid, hands-free mobile communication [1–20]. Many of these applications require the direction of an acoustic source in the presence of ambient noise, reverberation, interference and sensor noise. However, the accurate estimation of DOA at low signal-to-noise ratio (SNR) is very difficult. There are many different methods for DOA estimations including acoustic vector sensors and microphones array [21, 22]. The signals of microphones/hydrophones of uniform linear array (ULA) can be used to implement the spatial filtering for finding DOA of an acoustic source. There are numerous algorithms by which the DOA estimation of an acoustic source can be done using signals received at ULA such as time difference of arrival, beamforming, subspace, maximum likelihood, and compressed sensing. However, as SNR decreases the performance of these methods also degrades.

In recent years due to the technological advancement and easy availability of high-performance computing facilities, recurrent neural network (RNN) and support vector machine (SVM)-based techniques have been applied to various applications like speech recognition, speech separation. There are variety of methods by which SVM can be applied in the estimation of DOA estimation. DOA estimation problem can be solved using SVM by tackling it as a classification problem where SVM classifier will classify the signals at the microphones in different classes of DOA

S. Yadav (✉) · M. Wajid
Department of Electronics Engineering, ZHCET, Aligarh Muslim University,
Aligarh 202002, India
e-mail: mohdwajid@zhcet.ac.in

M. Usman
Department of Electrical Engineering, King Khalid University,
Abha 61411, Saudi Arabia
e-mail: musman@ieee.org; omfarooq@kku.edu.sa

S. Jain et al. (eds.), *Advances in Computational Intelligence Techniques*,
Algorithms for Intelligent Systems,
https://doi.org/10.1007/978-981-15-2620-6_18

angles. [23–25]. The work presented in this chapter is an extended and enhanced version of the work presented by Wajid et al. [32]. In this chapter, DOA for an acoustic source has been estimated using SVM model with ULA and results are compared with the delay and sum (DAS) beamforming and RNN model for different SNR [32].

The rest of the chapter is organised as follows. In Sect. 2, the signal model for ULA has been discussed. Section 3 gives a brief introduction of DAS beamforming, RNN and SVM. Simulation parameters and results are given in Sects. 4 and 5 conclude the chapter.

2 Signal Model

At a far-field (i.e. planar wave-front) a source has been placed, transmitting exponential narrowband signal $s(t)$ of wavelength λ from θ direction with respect to the vertical axis. A ULA receiver consisting of M number of microphones or hydrophones placed on a line aligned with the x-axis, and the separation between every two consecutive microphone/hydrophone is d. The signal received at the Mth microphone/hydrophone is given by

$$y_M(t) = s(t)e^{-j2\frac{\pi}{\lambda}(M-1)d\sin\theta} + n_M(t) \tag{1}$$

where $n_M(t)$ is noise added to the Mth microphone/hydrophone. Equation (1) can also be written as

$$\mathbf{y}(t) = [y_1(t), y_2(t), \ldots..y_M(t)]^T = \mathbf{a}(\theta)s(t) + \mathbf{n}(t). \tag{2}$$

where $\mathbf{a}(\theta)$ is the steering vector of the ULA, $\mathbf{n}(t)$ is the noise vector and $[.]^T$ denote transpose.

The $M \times M$ correlation matrix $\mathbf{C_{yy}}$ of received signal vector $\mathbf{y}(t)$ is expressed as

$$\mathbf{C_{yy}} = E\left[\mathbf{y}(t)\mathbf{y}^H(t)\right] = \mathbf{a}(\theta)\mathbf{S}\mathbf{a}^H(\theta) + \mathbf{C_n}, \tag{3}$$

where $[.]^H$ and $E[.]$ denote the conjugate transpose and ensemble average, respectively. Also the signal and noise correlation matrices $\mathbf{S}$ and $\mathbf{C_n}$ can be expressed as

$$\mathbf{S} = E\left[\mathbf{s}(t)\mathbf{s}^H(t)\right] \tag{4}$$

and

$$\mathbf{C_n} = E\left[\mathbf{n}(t)\mathbf{n}^H(t)\right], \tag{5}$$

respectively. Assuming, that all the noise components are mutually uncorrelated and have the same variance, therefore,

$$\mathbf{C_n} = \sigma^2\mathbf{I}, \tag{6}$$

where $\mathbf{I}$ is the identity matrix and σ^2 is the noise variance. Then, Eq. (3) can be expressed as

$$\mathbf{C_{yy}} = \mathbf{a}(\theta)\mathbf{S}\mathbf{a}^H(\theta) + \sigma^2\mathbf{I} \tag{7}$$

3 Methodology

3.1 DOA Estimation with Delay and Sum Beamforming

The DAS method computes the DOA by calculating signal power, $P(\phi)$ at each possible DOA and the estimated DOA is the argument of $P(\phi)$ at its maxima [26–28].

$$P(\phi) = \mathbf{a}(\phi)^H\mathbf{C_{yy}}\mathbf{a}(\phi) \tag{8}$$

where $\mathbf{a}(\phi)$ is the look direction vector in the direction ϕ.

3.2 DOA Estimation with Recurrent Neural Network

The DOA estimation problem of an acoustic source has been solved using RNN [29–32] where RNN based on bidirectional long short-term memory (BiLSTM) networks have been used to estimate the DOA. The RNNs are used where the time series or sequence of data has dependencies and BiLSTM is used for past and future dependencies in the data. The DOA estimation problem is being considered as a classification problem and for that BiLSTM network is being used. The block diagram of the BiLSTM-based RNN is shown in Fig. 1. The block diagram consists of an input layer for time series signal followed by a layer of BiLSTM. The DOA has been estimated by means of classification using a fully connected network followed by a softmax threshold. The details of bidirectional long short-term memory networks are given in [33].

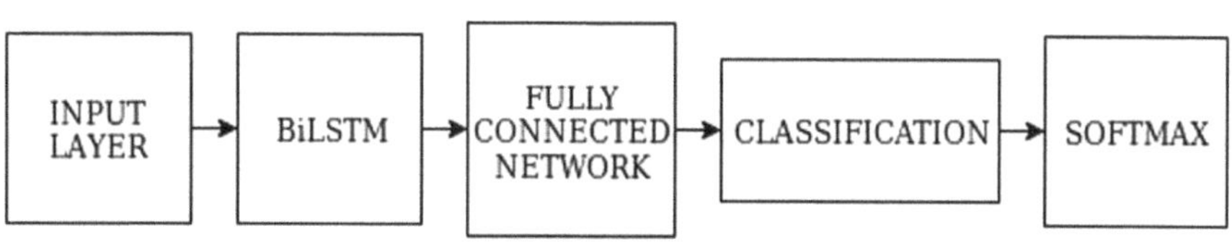

Fig. 1 Block diagram of RNN using BiLSTM for DOA estimation

3.3 DOA Estimation with Support Vector Machine Model

SVM as a supervised machine learning model can be used to combat the problem of DOA estimation. In this chapter, SVM based on one-vs-one multiclass support vector classifier [34] has been used for DOA estimation. Firstly, the input data (signals at the microphone/hydrophone) is pre-processed for proper feature extraction and then those features along with the labels are used for training the SVM classifier. SVM is a machine learning method for binary classification. The basic idea behind the support vector machine is to find a hyperplane to divide the data points in the N-dimensional plane into two regions representing the two classes for classification. In SVM, higher-dimensional space computation is not required, only the dot-product formula in that space is needed. The details of support vector machine are given in [35].

4 Simulation Parameters and Results

A uniform linear array consisting of four omnidirectional microphones is placed along the x-axis as shown in Fig. 2. The spacing between consecutive microphones d is kept equal to 10 cm. It has been assumed that the sound source is in a far-field emitting 1 kHz sinusoidal signal and speed of sound in air is considered to be 343 m/s. The relative attenuation of the signals impinging on the microphones/hydrophones has been neglected. All DOAs are estimated in clockwise direction with respect to the vertical axis. The received signals are of duration 25 ms and are sampled at 48 kHz. The training noisy signal vectors are generated from ULA's received signal vector after addition of noise with 26 dB SNR. For each DOA, 1400 independent noisy signal vectors are used for training. Testing has been performed on 600 independent noisy signal vectors for each DOA for each SNR value ranging from 26 to 6 dB with a constant decrement of 4 dB. There are four types of classification as given below

(a) *10-ary classification*: Range of DOA is from 0° to 90° with a constant increment of 10°.
(b) *19-ary classification*: Range of DOA is from 0° to 90° with a constant increment of 5°.
(c) *46-ary classification*: Range of DOA is from 0° to 90° with a constant increment of 2°.
(d) *91-ary classification*: Range of DOA is from 0° to 90° with a constant increment of 1°.

The values of the above parameters are taken to be the same as in [32], so that the results can be compared.

In the SVM model being used here, the kernel type is $linear$ and the regularization parameter C is set to 1. As the input data cannot be directly fitted into the support vector classifier of SVM, proper features are extracted from the data. Here, cross-correlation coefficients between every unique pair of signals out of four signals

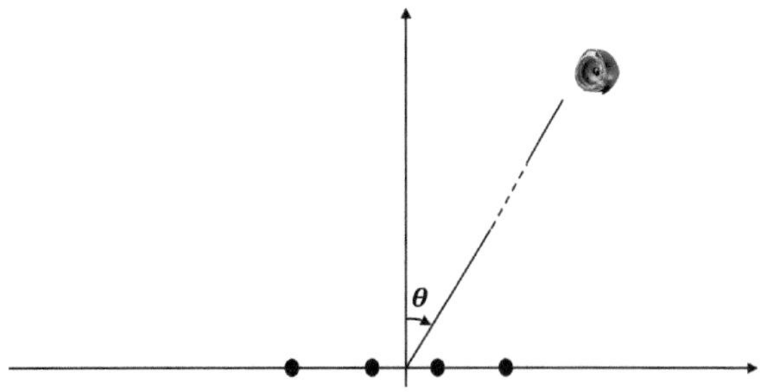

Fig. 2 Diagram of uniform linear array. Where circles indicate the microphones separated by distance of 10 cm and speaker is in the far-field at an angle of θ with respect to vertical axis in the clockwise direction

($^4C_2 = 6$) at microphones are being used as the features of this SVM classifier model. As there are six features, data points will be marked in the six-dimensional space and hyperplanes generated by SVM will separate data points between every pair of classes. In this way, binary classifier of SVM can be extended for multiclass classification.

The performance of the trained network and support vector classifier has been evaluated in terms of accuracy A for each true angle class θ which is defined as

$$A(\theta) = \frac{\text{Correct number of predictions of angle } \theta}{\text{Total number of predictions of angle } \theta} \tag{9}$$

and average accuracy $\bar{A}$ is the average of accuracy for all DOAs as given below

$$\bar{A} = \frac{1}{\text{NoC}} \sum_{\forall \theta} A(\theta). \tag{10}$$

where NoC is the number of classes.

Figures 3, 4, 5 and 6 show the testing accuracy for n-ary (n = 10, 19, 46, and 91) class testing and its comparison with the DAS beamforming. For the DAS as well as RNN, the average accuracy decreases with the decrease in SNR, also the accuracy is higher at the broadside (0°) than at the end-fire (90°). The DOA estimation accuracy decreases with the increase in DOA (moving from broadside to end-fire). However, SVM has very poor accuracy at both the broadside and the end-fire but fairly large accuracy at middle points. SVM performs better in 46-ary and 91-ary classification because it gives higher accuracy even at low value of SNR. Also, the RNN-based results have better accuracy than the DAS beamformer for all values of SNR.

Figure 7 gives the confusion matrix for 10-ary classification at 6 dB SNR using SVM model. Some of the off-diagonal element of the matrix for SVM model is non-zero for the target class of 0° and 90°; however, the classifications of these target classes are adjacent to the true classes. From the confusion matrix it can be seen that for all the angles except two extremes, the accuracy is 100%. But, off-diagonal elements are non-zero for 60°, 70°, 80° and 90° in case of RNN model but still the maximum value lies in the diagonal elements.

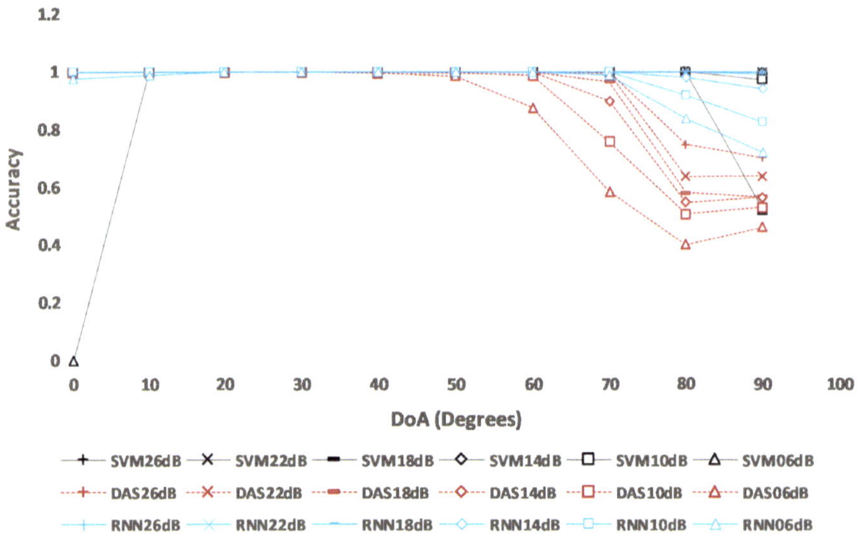

(a) Accuracy versus DOA (degrees)

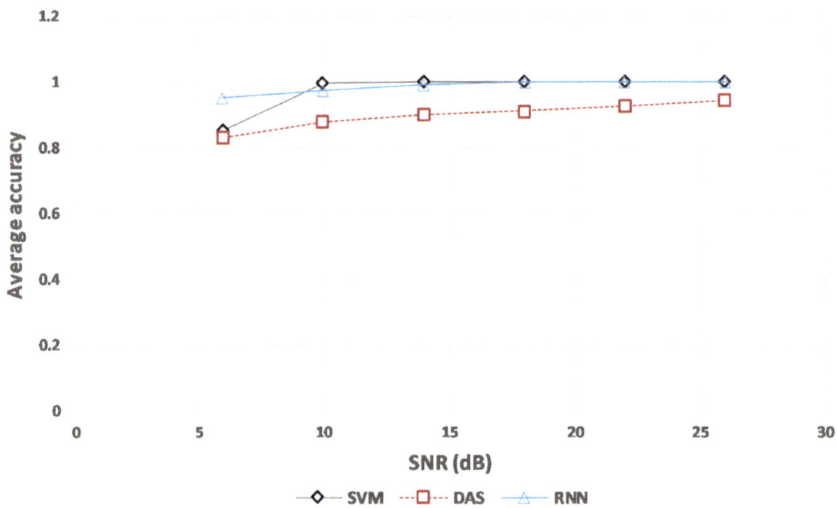

(b) Average accuracy versus SNR (dB)

Fig. 3 Results for 10-ary class testing with 600 noisy signal vectors for each value of SNR using SVM and RNN models and there comparison with the results of DAS beamforming

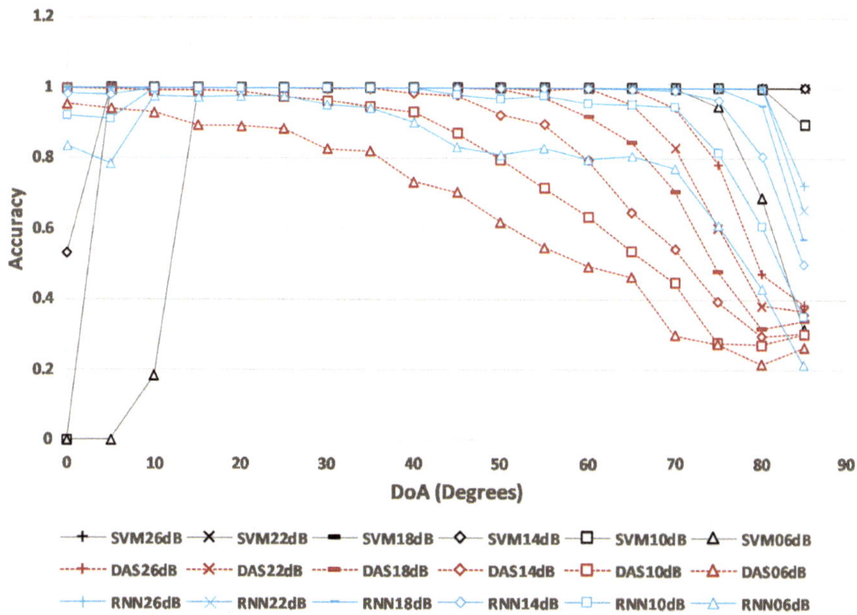

(a) Accuracy versus DOA (degrees)

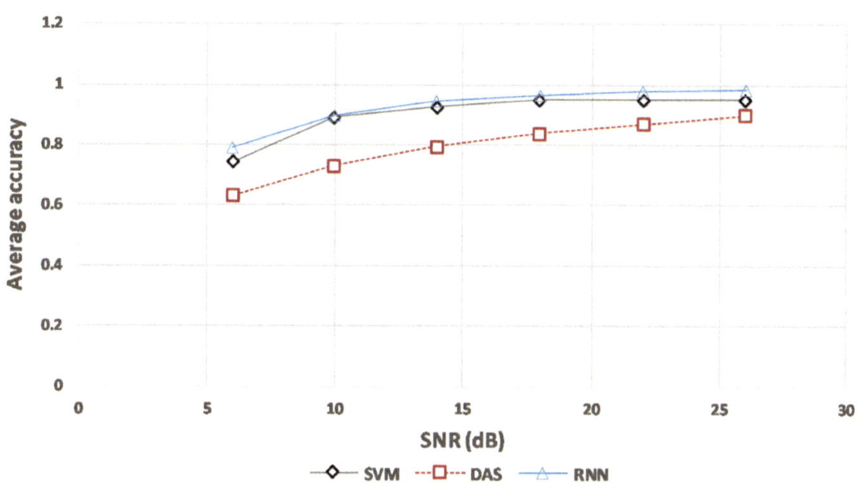

(b) Average accuracy versus SNR (dB)

Fig. 4 Results for 19-ary class testing with 600 noisy signal vectors for each value of SNR using SVM and RNN models and there comparison with the results of DAS beamforming

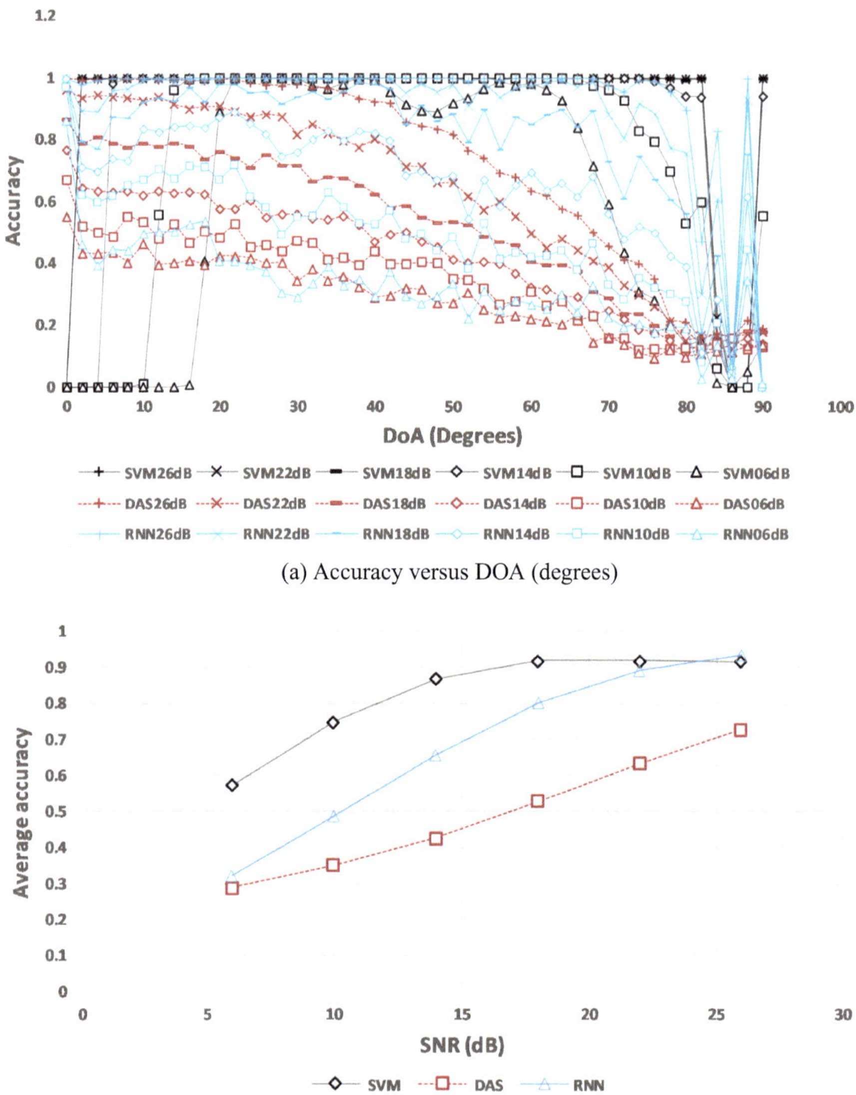

(a) Accuracy versus DOA (degrees)

(b) Average accuracy versus SNR (dB)

Fig. 5 Results for 46-ary class testing with 600 noisy signal vectors for each value of SNR using SVM and RNN models and there comparison with the results of DAS beamforming

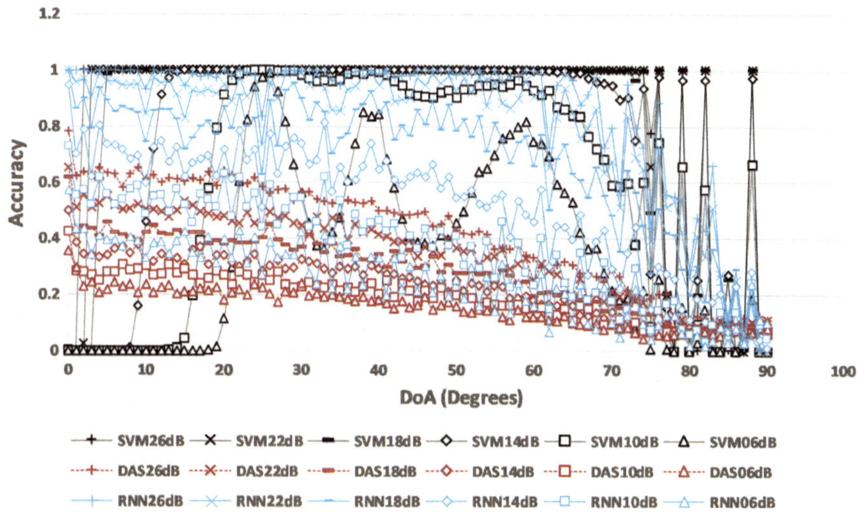

(a) Accuracy versus DOA (degrees)

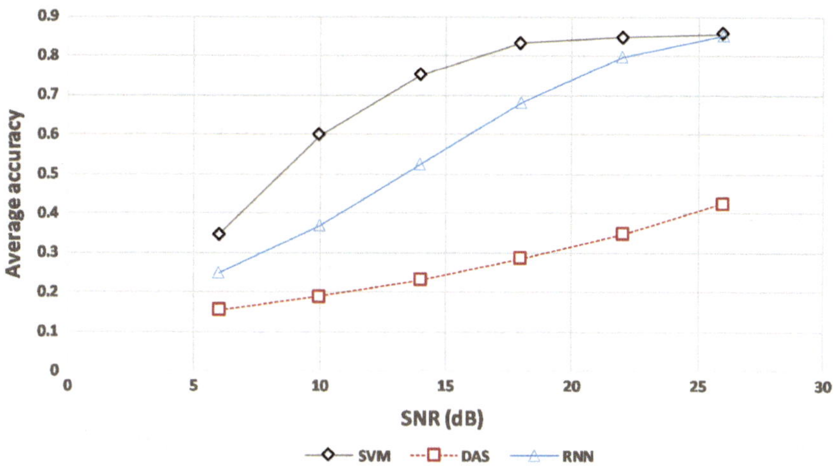

(b) Average accuracy versus SNR (dB)

Fig. 6 Results for 91-ary class testing with 600 noisy signal vectors for each value of SNR using SVM and RNN models and there comparison with the results of DAS beamforming

5 Conclusion

It has been concluded that performance of SVM-based DOA estimation is better than DAS beamforming as well as RNN. The performance of SVM is very close to that of RNN for higher SNR but when the SNR is very low, SVM outperforms RNN. However, SVM has very low accuracy at both extremes of DOA angles and very high

Output Class \ Target Class	0°		10°		20°		30°		40°		50°		60°		70°		80°		90°	
0°	100	0	0	0	0	0	0	0	0	0	0	0	0	0	0	0	0	0	0	0
10°	0	100	100	100	0	0	0	0	0	0	0	0	0	0	0	0	0	0	0	0
20°	0	0	0	0	100	100	0	0	0	0	0	0	0	0	0	0	0	0	0	0
30°	0	0	0	0	0	0	100	100	0	0	0	0	0	0	0	0	0	0	0	0
40°	0	0	0	0	0	0	0	0	100	100	0	0	0	0	0	0	0	0	0	0
50°	0	0	0	0	0	0	0	0	0	0	100	100	0	0	0	0	0	0	0	0
60°	0	0	0	0	0	0	0	0	0	0	0	0	99.6	100	0	0	0	0	0	0
70°	0	0	0	0	0	0	0	0	0	0	0	0	0.33	0	97.6	100	1.3	0	0.16	0
80°	0	0	0	0	0	0	0	0	0	0	0	0	0	0	2.3	0	63.6	100	18.3	47.7
90°	0	0	0	0	0	0	0	0	0	0	0	0	0	0	0	0	35	0	81.5	52.2

Fig. 7 Confusion Matrix for 10-ary classification at 6 dB SNR with first entry for RNN and second entry for SVM (in %)

accuracy at the intermediate DOA angles while RNN has uniform accuracy over the entire range of angles except the end-fire (90°). When limited amount of data is used for training, then feature-based supervised machine learning models perform better as compared to neural networks. Another reason for better performance in the case of SVM is the cross-correlation operation used in feature extraction which results in suppression of noise and hence gives better accuracy.

References

1. Zheng X, Ritz C, Xi J (2016) Encoding and communicating navigable speech soundfields. Multimed Tools Appl 75(9):5183–5204
2. Asaei A, Taghizadeh MJ, Saeid H, Raj B, Bourlard H, Cevher V (2016) Binary sparse coding of convolutive mixtures for sound localization and separation via spatialization. IEEE Trans Signal Process 64(3):567–579
3. Bekkerman I, Tabrikian J (2006) Target detection and localization using mimo radars and sonars. IEEE Trans Signal Process 54(10):3873–3883
4. Wong KT, Zoltowski MD (1997) Closed-form underwater acoustic direction-finding with arbitrarily spaced vector hydrophones at unknown locations. IEEE J Ocean Eng 22(3):566–575
5. Sheng X, Hu YH (2005) Maximum likelihood multiple-source localization using acoustic energy measurements with wireless sensor networks. IEEE Trans Signal Process 53(1):44–53
6. Zhao S, Ahmed S, Liang Y, Rupnow K, Chen D, Jones DL (2012) A real-time 3d sound localization system with miniature microphone array for virtual reality. In: 2012 7th IEEE conference on industrial electronics and applications (ICIEA). IEEE, pp 1853–1857
7. Clark JA, Tarasek G (2006) Localization of radiating sources along the hull of a submarine using a vector sensor array. In: OCEANS 2006. IEEE(OCEANS), pp 1–3

8. Carpenter RN, Cray BA, Levine ER (2006) Broadband ocean acoustic (boa) laboratory in narragansett bay: preliminary in situ harbor security measurements. In: Defense and security symposium. International Society for Optics and Photonics, pp 620409–620409
9. DiBiase JH, Silverman HF, Brandstein MS (2001) Robust localization in reverberant rooms. In: Microphone arrays. Springer, pp 157–180
10. Bechler D, Schlosser MS, Kroschel K (2004) System for robust 3d speaker tracking using microphone array measurements. In: Proceedings. 2004 IEEE/RSJ international conference on intelligent robots and systems (IROS 2004), vol 3. IEEE, pp 2117–21222004
11. Argentieri S, Danes, P (2007) Broadband variations of the music high-resolution method for sound source localization in robotics. In: 2007. IROS 2007. IEEE/RSJ international conference on intelligent robots and systems. IEEE, pp 2009–2014
12. Nakadai K, Matsuura D, Okuno HG, Kitano H (2003) Applying scattering theory to robot audition system: Robust sound source localization and extraction. In: Proceedings. 2003 IEEE/RSJ international conference on intelligent robots and systems (IROS 2003), vol 2. IEEE, pp 1147–1152
13. Zhao S, Chng ES, Hieu NT, Li H (2010) A robust real-time sound source localization system for olivia robot. In: 2010 APSIPA annual summit and conference
14. Xiao X, Zhao S, Nguyen DHH, Zhong X, Jones DL, Chng ES, Li H (2011) The ntu-adsc systems for reverberation challenge 2014. In: Proc. REVERB challenge workshop
15. Delikaris-Manias S, Vilkamo J, Pulkki V (2016) Signal-dependent spatial filtering based on weighted-orthogonal beamformers in the spherical harmonic domain. IEEE/ACM Trans Audio, Speech Lang Process (TASLP) 24(9):1507–1519
16. Delikaris-Manias S, Pulkki V (2013) Cross pattern coherence algorithm for spatial filtering applications utilizing microphone arrays. IEEE Trans Audio, Speech, Lang Process 21(11):2356–2367
17. Zhang C, Florêncio D, Ba DE, Zhang Z (2008) Maximum likelihood sound source localization and beamforming for directional microphone arrays in distributed meetings. IEEE Trans Multimed 10(3):538–548
18. Van den Bogaert T, Carette E, Wouters J (2011) Sound source localization using hearing aids with microphones placed behind-the-ear, in-the-canal, and in-the-pinna. Int J Audiol 50(3):164–176
19. Widrow B (2000) A microphone array for hearing aids. In: Adaptive systems for signal processing, communications, and control symposium 2000. AS-SPCC. The IEEE 2000. IEEE, pp 7–11
20. Abdi A, Guo H, Sutthiwan P (2007) A new vector sensor receiver for underwater acoustic communication. In: *OCEANS 2007*. IEEE, pp 1–10
21. Wajid M, Kumar A, Bahl R (2017) Direction-of-arrival estimation algorithms using single acoustic vector-sensor. In: 2017 international conference on multimedia, signal processing and communication technologies (IMPACT). IEEE, pp 84–88
22. Wajid M, Kumar A, Bahl R (2017) Direction-finding accuracy of an air acoustic vector sensor in correlated noise field. In: 2017 4th international conference on signal processing, computing and control (ISPCC). IEEE, pp 21–25
23. Donelli M, Viani F, Rocca P, Massa A (2009) An innovative multiresolution approach for doa estimation based on a support vector classification. IEEE Trans Antennas Propag 57(8):2279–2292
24. Lizzi L, Oliveri G, Rocca P, Massa A (2010) Estimation of the directions-of-arrival of correlated signals by means of a svm-based multi-resolution approach. In: 2010 IEEE antennas and propagation society international symposium. IEEE, pp 1–4
25. Rohwer JA, Abdallah CT, Christodoulou CG (2003) Least squares support vector machines for direction of arrival estimation. In: IEEE antennas and propagation society international symposium. Digest. Held in conjunction with: USNC/CNC/URSI North American Radio Sci. Meeting (Cat. No. 03CH37450), vol 1. IEEE, pp 57–60
26. Van Veen BD, Buckley KM (1988) Beamforming: a versatile approach to spatial filtering. IEEE Assp Mag 5(2):4–24

27. Haykin S (1985) Array signal processing. Englewood Cliffs, NJ, Prentice-Hall, Inc., 1985, 493 p. For individual items see A85-43961 to A85-43963
28. Manolakis DG, Ingle VK, Kogon SM et al (2000) Statistical and adaptive signal processing: spectral estimation, signal modeling, adaptive filtering, and array processing. McGraw-Hill Boston
29. Kase Y, Nishimura T, Ohgane T, Ogawa Y, Kitayama D, Kishiyama Y (2018) DOA estimation of two targets with deep learning. In: 2018 15th workshop on positioning, navigation and communications (WPNC). IEEE, pp 1–5
30. Li Q, Zhang X, Li H (2018) Online direction of arrival estimation based on deep learning. In: 2018 IEEE international conference on acoustics, speech and signal processing (ICASSP). IEEE, pp 2616–2620
31. Liu Z-M, Zhang C, Philip SY (2018) Direction-of-arrival estimation based on deep neural networks with robustness to array imperfections. IEEE Trans Antennas Propag 66(12):7315–7327
32. Wajid M, Kumar B, Goel A, Kumar A, Bahl R (2019) Direction of arrival estimation with uniform linear array based on recurrent neural network. In: 5th international conference on signal processing, computing and control (ISPCC). IEEE, p 17
33. Hochreiter S, Schmidhuber J (1997) Long short-term memory. Neural Comput 9(8):1735–1780
34. Debnath R, Takahide N, Takahashi H (2003) A fast learning decision-based SVM for multi-class problems. In: ICMLA, pp 128–134
35. Boswell D (2002) Introduction to support vector machines. Departement of Computer Science and Engineering University of California San Diego

Author Index

S. Jain et al. (eds.), *Advances in Computational Intelligence Techniques*,
Algorithms for Intelligent Systems,
https://doi.org/10.1007/978-981-15-2620-6